GILDED CAGE

Other bestsellers by the author

Princestan: How Nehru, Patel and Mountbatten Made India

Bonfire of Kashmiriyat: Deconstructing the Accession

GILDED CAGE

YEARS THAT MADE AND UNMADE KASHMIR

SANDEEP BAMZAI

RUPA

Published by
Rupa Publications India Pvt. Ltd 2023
7/16, Ansari Road, Daryaganj
New Delhi 110002

Sales centres:
Allahabad Bengaluru Chennai
Hyderabad Jaipur Kathmandu
Kolkata Mumbai

P-ISBN: 978-93-5520-812-5
E-ISBN: 978-93-5520-814-9

First impression 2023

10 9 8 7 6 5 4 3 2 1

Printed in India

CONTENTS

FOREWORD

'BAPUJI WAS A PATRIOT AND A NATIONALIST'

It began when Viceroy Lord Mountbatten and Vicereine Edwina visited Srinagar to meet with Bapuji (Maharaja Hari Singh). There was a big reception laid out for him at Gulab Bhawan, with great flourish and drama. Before that, there were a series of garden parties, banquets and receptions. During the dinner, Mountbatten accidentally pressed the bell under the gigantic table, which created great consternation, with all the diners alarmed. This bell was supposed to be pressed by Hari Singh after the formal banquet was over. Given that Mountbatten was tall, his knee had mistakenly pressed the bell, and the band outside started playing 'God Save the King', for which everyone had to stand midway during the chicken curry. Bapuji was furious, as he was a stickler for protocol, and this was to happen to close the dinner. Mountbatten laughed uproariously, making nothing of it. I, too, was in a fit of giggles despite my mother's glares. This wasn't all—my father had to introduce all his courtiers to Lord Mountbatten at an earlier garden party, and there, too, there had been chaos, for he had memorized the names, but the men had been standing in a different order, leading to Babuji getting red in the face. Remember, those were difficult times—there was a sense of unease across the state. The deadline to sign in favour of one of the two dominions was fast approaching. I suspect my father still believed that the British would not leave India. Indecisive by nature, he played for time. A typical feudal reaction to a difficult situation is to avoid facing it, and my father was particularly prone to resort to this.

Vacillating, Father refused to see him and instead packed him off to Thricker fishing lodge, where, in his boredom, the Viceroy swam buck naked. On his return, too, Bapuji avoided meeting him, saying he had colic. Personally, I think he should have met Mountbatten—after all, he was visiting in his capacity as viceroy. This would have helped father make up his mind and expedite the accession to the dominion of India. It was an opportunity missed—the last real chance of working out a viable political settlement was lost. Let us also understand his predicament a little better: it was complex—if he acceded to Pakistan, a large chunk of his people, including his entire Dogra base, would have been outraged. If he acceded to India, he risked alienating a large section of his Muslim subjects.

Earlier, the Congress president of the time, J.B. Kripalani, came to meet Father and visited Hari Niwas in Jammu. I do not remember what transpired in that meeting, but I am aware that nothing concrete fructified because Kripalani went back disappointed.

This was followed by a visit from Gandhiji. Here, again,

my father was reluctant to meet him, but my mother, Maharani Tara Devi, who was much closer to reality than my father, told him that he was no ordinary man, it was Mahatma Gandhi himself who had come calling, and it would be downright rude not to meet him. In fact, I, too, wanted to be present for the same reason—to catch a glimpse of the Mahatma—so I came in my wheelchair. Father was completely attentive and respectful, but Gandhiji spoke softly, so I couldn't fathom what was being said. However, when I was introduced, Gandhiji asked: '*Kaise ho* (How are you)?' I understood one thing from the conversation—Gandhiji clearly told Babuji to ascertain the will of the people of the state and find out from them which way they wanted to go. The meeting took place under one of the chinar trees in the front lawn of Gulab Bhavan. Special arrangements were made for goat milk and fruits, we took our places, and at the appointed hour of 5.00 p.m., 1 August, he arrived. After an hour and a half, he got up to go.

Guru Golwalkar, the founder of the RSS (Rashtriya Swayamsevak Sangh), also visited Kashmir and met my father, and while I got a glimpse of the man, I have no recollection of what happened between him and my father.

Around this time, there was a lot of toing and froing, and two Pathans, I remember, from (Abdul) Ghaffar Khan's family also came to meet my father. Those were horrible times, very tense; and in parallel, I could see from my prism in the palace, my mamaji, Thakur Nachint Chand, was actively playing the role of interlocutor between Sheikh Abdullah and my father. The prime minister, who was also my guardian and who is someone painted negatively but had great poise and intelligence, Ram Chandra Kak, was also a key member of the court. His wife, Margaret, and he are much-maligned figures in our history, but he had the intellect to pull off impossible things. My father was fond of him, and as maharaja, I remember him going to Kak's brother Amarnath Kak's daughter's wedding. Kak's single-biggest error in judgement was to arrest Pandit Nehru in Domel when he came to play the role of Sheikh's lawyer in the sedition trial. It was a mistake of epic proportions—he arrested Jawaharlal Nehru, an unforgettable decision that caused much pain.

One of the most powerful figures of the time was Swami Sant Deo, a pro-RSS shadowy figure who constantly misled my father. He would be whispering in Bapuji's ear: '*Aapka jhanda Lahore pe chadega, Kashmir phir se raj karega* (Your flag will be raised in Lahore, Kashmir will rule again).' He was one of the many who planted the seed of remaining independent in my father's head. The crown jewel (Kashmir State), which had borders with big powers of the time, wedged as it was between so many nations, could hardly survive on its own. Strangely, I had no communication with him, but he peddled a lot of influence in the court. My father would tell me that the British would never leave India and that decolonization was a myth. He never realized that change would dawn on all of us so rapidly. My father was an extremely intelligent man, but the historical forces that were in play were far greater than what he could grasp. There were four major forces—a rising Congress, led by an aggressive Nehru and Sardar Patel; the Muslim League under Jinnah, who wanted to poach Kashmir at any cost; an anti-Muslim

League Muslim-majority party led by Sheikh Abdullah, who was equally viscerally anti-Dogra; and, of course, the British, who didn't want to leave, and their Political Department was busy playing its own games.

All this changed frenetically. The epicentre of activity was the lower corridor of the eastern wing of the palace, where everyone gathered. My father's friend Victor Rosenthal used to try cheering up Babuji; the Maharani and courtiers would all be present. Swami Sant Deo had, by then, retreated to his Chasma Shahi Guest House, but as the situation deteriorated, his visits became infrequent, and my father's interest in him began to decrease. From chief counsel, I reckon, my father came to realize that the Swami's pretensions to great occult powers were somewhat overrated and his old skepticism had begun to reassert itself.

My father, Maharaja Hari Singh, was first and foremost a patriot and a nationalist. It is a miracle that we survived despite gale-force conditions and all these forces at work in the theatre called Kashmir. Nobody wanted to let go of it—not Jinnah, not Nehru, not the British, not Sheikh and perhaps not even my father, who dreamt of an independent eastern Switzerland. The anti-colonial movement was spreading like a bush fire, not just in India but, post the Second World War, it was taking root across colonies as they tried overthrowing the yoke of lifelong tyranny. In these tenuous circumstances, my father wrote to Mountbatten on 26 October, acceding to the dominion of India. He wrote: 'I wanted to take time to decide to which Dominion to accede...or stand independent.' V.P. Menon asked us to leave Srinagar as the raiders were closing in and had burnt down the Mahura power station. Bapuji drove from Srinagar to Jammu, himself leading the convoy of vehicles, and by his side was his friend Viktor Rosenthal. When we reached Jammu, he said, 'We have lost Kashmir.' He looked forlorn and a tragic figure. As a ruler, he had been progressive, although I must add here that Kashmiri Pandit historians have been grossly unfair to him.

Although Kashmiri Pandits came into their own after centuries of Mughal, Afghan and Sikh rule, it is only under Dogra rule that they flourished, reaching even the prime ministership. I am always surprised that in all the books I have read on Kashmiri Pandits, writers have never given any credit to the Dogras. In fact, the creation of the Jammu and Kashmir state by Maharaja Gulab Singh in the nineteenth century was a major geopolitical event, which has not received the attention that it deserves. Personally, let me say that I owe a deep debt to the Kashmiri Pandit community, who were my tutors and guardians for many years. I would like to mention Prof. B.K. Madan, Shri R.C. Kak, Col. K.N. Haksar, Prof. P.N. Chaku, Prof. J.N. Bhan, Pandit Parmanand and, of course, the great Swami Lakshmanjoo.

On that fateful day, I remember distinctly, the darbar was taking place at the City Palace on the Jhelum when the lights went out and panic set in as the Afridis raced headlong down the road from Mahura. Our smug world had collapsed around us—the wheels of destiny had turned full circle. A flurry of activity began—Menon's air dash to convince Bapuji to leave Srinagar, the signing of the Instrument of Accession and then the midnight race to Jammu. At first light, still 60 miles from Jammu, a cream-coloured car joined our grim procession. Yes, the Swami Sant Deo had arrived, obviously his miraculous powers did not include the capacity to face the raiders.

Once we moved to Jammu, the Maharaja of Patiala came with his fauj; the army commander Brig. Paranjpe was a frequent visitor; and even Pandit Nehru came (and my father brought him to my room). Sardar Patel came, and this was the time I was still in a wheelchair. He quizzically asked why I was in a wheelchair and insisted that I be sent to the US for treatment. Almost overnight, arrangements were made, and I landed in New York on 31 December and stayed there for the whole of 1948. After that, I was totally out of the mix, but my father kept me abreast of developments by writing regularly to me. I had left Kashmir, but Kashmir was inside me. Both Sheikh Abdullah and Sir Gopalaswami Ayyangar met me in NYC. When I returned in 1949, the tension was palpable and intolerable. Sheikh was chief administrator, while Pathania was the diwan—things were and felt different. The situation seemed to be careening out of control, as the tension between my father and Sheikh had reached a breaking point.

Then it appeared that the Government of India had taken a view on the matter, and we were called to the Imperial Hotel in Delhi, where Sardar Patel met us. He asked Bapuji to stay out of the state for six months, didn't call it an abdication, while the Maharani, who was people-friendly, was sent to Kasauli. On 20 June 1949, I was appointed regent. Bapuji was most upset with the whole thing, more so because he was banished to Bombay, but he never expressed any bitterness. Panditji, more or less, adopted me and became my mentor at this juncture. The conflict between my father and me arose over my appointment as Sadr-e-Riyasat. Sheikh proclaimed that monarchy had been abolished, and the Dogras were furious with me, for I followed Nehru and not my own father. Remember that I was 21 years old and my dilemma was serious: had I chosen the right path? My father was justifiably furious at the way he had been treated. The Dogras perceived my acceptance of the regent's position as a traitorous act, which would lend an aura of legitimacy to Sheikh's actions—Sheikh's politics being predicated on abolishment of the Dogra dynasty and my father's abdication.

I was invited for breakfast by Panditji, where I was given a historical perspective to what had transpired. I was told that I would be part of a grand transition from feudalism to democracy. I accepted this, but on my first visit to Jammu thereafter, black flags greeted me. Those were trying times. Sheikh Abdullah tried to implicate my father in the Jammu riots, which was grossly unfair. It is ironic that, as Sadr-e-Riyasat, it was my turn to dismiss him in 1953, after which I became a hero again for the Dogras. On the morning of his arrest from Gulmarg, I invited Sheikh Abdullah to my residence to have a chat. The conversation revealed his true state of mind—from a position of clearly endorsing the accession to India as final and irrevocable, he had, over the past few months, moved into an entirely different posture. He was evidently expecting some sort of international pressure to be brought upon India to grant virtual independent status to Kashmir, and this was quite obviously the reason behind holding up the implementation of the Delhi Agreement, which would further cement the relationship between the state and the Centre. In our meeting, he was not only unable to give me any assurance that the internal crisis in his Cabinet would be resolved but had added a new dimension of uncertainty by talking about an external solution. Our meeting lasted 45 minutes. That same night, after it was decided that he would dismissed, he flew into a rage on seeing and reading the warrant signed by me. He shouted, 'Who is the Sadr-e-Riyasat to dismiss me? I made the chit of a boy Sadr-e-Riyasat.' Arrogance emanating from possession of power leading to personal aggrandisement had eluded him.

Many books have been written on my father's decision to accede to India on 26 October 1947. Sandeep has his own reading and analysis of the situation, with all of which we may not agree. However, I commend him for having devoted his time and energy to this eternally fascinating event in contemporary Indian history. My text is really in the nature of a conversation that I had with Sandeep Bamzai, which he has decided to use as a foreword. I have known his family for three generations and am impressed by his keen interest in our home state.

Dr Karan Singh

Former Sadr-e-Riyasat (1958–65) and
governor (1965–67) of Jammu and Kashmir

PREFACE

The Kashmir story has been nothing but as dramatic as the tale of its accession. In an age of right to information, what better than to try and piece together this story of Kashmir, as it has never been told before. A catalogue of writing, both domestic and international, has tried to dissect and disseminate the Kashmir problem, some successfully and others not so successfully. While Pakistan now labels it as an indigenous uprising or they prefer to give it the nomenclature of a 'freedom struggle', the reality is that the seeds of the Kashmir imbroglio were sown in the dark days of the accession.

I have deliberately chosen the period between 1931 and 1953 when the downtrodden populace of the Valley rose against the representative of the repressive rule of the Dogra dynasty—Maharaja Hari Singh. They were led by a man who was, for all practical purposes, a nationalist, though disturbing evidence over time points otherwise—Sheikh Abdullah, a nationalist who somewhere along the way got boxed in due to the contradictions that traversed the corridors of his mind. In his mind, he could never quite define internal autonomy for Kashmir, and as such, this acquired the hue of independence from the Union of India. This story shows how his thinking evolved across that vector—from nationalist to preacher and harbinger of autonomy for the Valley. For that reason alone I chose to end the book with his controversial arrest in August 1953.

But this story is as much about Pandit Jawaharlal Nehru, that titan amongst statesmen, and his friendship with Abdullah. It is also about Nehru's resolve stoked by his private secretary Dwarkanath Kachru's (another Kashmiri) inputs. Nehru was convinced that Kashmir was worth fighting for, and he advocated the Mysore Model after its accession. In Mysore, the leader of the popular party was asked to choose his colleagues, he himself being the prime minister. His hope sprung from the fact that Sheikh Abdullah, a Muslim leader would lead the state out of the morass that it had sunk into during the Maharaja's rule. Economic aspirations and all-inclusive growth were the bulwarks of the New Kashmir movement, and this was in many ways Sheikh's mandate.

Intertwined in this tale is the larger story of the Indian princely order, of which Maharaja Hari Singh was very much a part—their hopes and ambitions in the run up to Independence and the lapse of paramountcy. They are equally a part of the drama that was unfolding in the days preceding Partition. And finally, the story dovetails the role of Sardar Vallabhbhai Patel, Pandit Nehru's disparate twin as it were in the freedom struggle and formation of India's first government. Facilitator of the accession, antithesis of Sheikh, supporter of Hari Singh, acting as a counterweight to the Nehru–Sheikh axis, Patel stood tall against the princes and their collective and individual intrigues and machinations as they tried to stay out of the ambit of the burgeoning might of the Congress party.

It is also about Mohammed Ali Jinnah, a man who tried his level best till the day he died to grab Kashmir by hook or by crook. A man who visited Kashmir for holidays in the late 1920s and then again in 1936 till he finally arrived for what many describe to a hero's welcome in May 1944. This visit, in response to an invitation extended by Muslim Conference leaders, saw him spending over two months in Kashmir—a semi-official visit where he was deliberately shunned by the ruler and his prime minister. Jinnah had come to Kashmir to effectively unify the Muslim Conference, which owed its allegiance to the Muslim League and Sheikh Abdullah-led National Conference, but this did not fructify, as Jinnah could never convince Abdullah, who soon after the former left the Valley, attacked him in a series of mass contact meetings. The differences of opinion between Jinnah and Abdullah well and truly cost the former—Kashmir. And remember, Sheikh Sahab had Nehru, and Nehru had Abdullah in Kashmir. And together, they thwarted Jinnah's subterfuge. Coercive tactics were used by Jinnah subsequently to strong-arm the accession, but this did not materialize, making Kashmir the unfinished business of Partition for every successive regime in Pakistan.

Equally, this treatise is about my grandfather, K.N. Bamzai, a journalist with *Blitz* who, like many of that age, was besotted by what Nehru represented. First as chief of bureau of *Blitz* in Delhi, scooping story after story, then as OSD to Nehru, followed by personal secretary to Jammu and

Kashmir's first prime minister, Sheikh Abdullah, and then OSD again on Kashmir to Nehru in Delhi for long years, my grandfather always had a ringside view of historic events in India's capital, and with an astute finger on the pulse on Kashmir affairs, he always had skin in the game. His vast repository of secret and confidential papers, missives, documents, aides-memoires have helped me reconstruct the third and final part of my trilogy on my *madre vatan*, Kashmir. I have deep dived into history over these last four years to write two definitive and contrasting tales emerging out of the same fountainhead of India's brutal partition, a result of Britain's decolonization.

PROLOGUE

THE COUP THAT FAILED

In the white heat of a bruising battle, the capacity to think and act is lost. Delusion takes over, but real time tells you that there is mystery, deception, allegory and even chicanery, yet, what you see is what you get. True belief is held in suspended animation, the feeling that you have the trigger pressed, target scoped and yet are incapacitated to fire taking over. Plagued by doubts, you wonder whether things will fructify the way you planned them.

On 24 October 1947, the maharaja of Jammu and Kashmir (J&K), Hari Singh, made an urgent appeal to the Indian government. His kingdom was in the throes of an invasion, and he was running out of options. While he waited for an answer, the Indian Cabinet's Defence Committee met in Delhi to take stock of the situation. V.P. Menon, administrative head and secretary of the state department under Sardar Vallabhbhai Patel, was instructed to fly to Srinagar on 25 October. Menon's first priority was to get the Maharaja and his family out of Srinagar. There were no forces left to guard the capital, and the invaders were virtually knocking at the door.

On 25 October, Menon arrived at Srinagar to study the situation. He met with Justice Meher Chand Mahajan—prime minister (PM) of Kashmir at the time—and sent the Maharaja from Srinagar to Jammu, as there was news that the raiders had started infiltrating Srinagar.

Menon returned to New Delhi in the morning of 26 October and brought Mahajan with himself. Menon reported to Jawaharlal Nehru and Patel that the Maharaja 'had gone to pieces completely' and could 'come to no decision' regarding the future of his kingdom. Mahajan recalls: 'I requested immediate military aid on any terms... Give us the military force we need. Take the accession and give whatever power you desire to the popular party. The army must fly to save Srinagar this evening or else I will go to Lahore and negotiate terms with Mr Jinnah.'[1] Nehru was furious at hearing this.

After the meeting, the government decided to fly two companies of troops to Srinagar on the morning of 27 October. Menon himself took a plane to Jammu, where the Maharaja was stationed. Governor-General Mountbatten had contended that it would be the 'height of folly' to send troops to a neutral state without an accession completed and 'that it should only be temporary prior to a referendum.'[2] Menon carried a message for the Maharaja: he had to join the Union if he wanted to ward off the invasion. The Maharaja was ready to accede. In fact, according to Menon's

---

[1]Mahajan, Mehr Chand, *Looking Back: The Autobiography of Mehr Chand Mahajan*, Asia Publishing House, Delhi, 1963.

[2]Kaul, Santosh, et. al. (eds), 'Document 116' *Kashmir: Legal Documents*. Light & Life Publishers, 1977.

memoirs, the Maharaja had left word with an aide that if Menon did not return with an offer, he was to shoot the Maharaja in his sleep.[3] Hari Singh signed the accession letter, regretting that the invasion had left him with no time to decide what was in the best interest of his state—to stay independent or merge with India or Pakistan. Menon took the documents immediately back to Delhi, where he met Patel at the aerodrome, who then went straight to the meeting of the Defence Committee, which finally decided to accede Kashmir to India, subject to a plebiscite later on.

Before leaving for Jammu on October 27, Mahajan had asked Nehru to give in his own handwriting the conditions on which the Maharaja would be given military help. Nehru wrote asking for accession on three subjects—first, jurisdiction over matters of defence, external affairs and transport. Second, internal administration of the state to be democratized and a new constitution framed on the lines of the model set out for the state of Mysore. The third decree was that Sheikh Abdullah should be included in the state administration set-up. These were shown to the Maharaja, who accepted all three conditions.

Two of the principal characters in this thrilling chase to retain Kashmir were Pandit Nehru and Sheikh Abdullah. In their joint quest for Kashmir, they were constantly thwarted by the man who created Pakistan—Mohammed Ali Jinnah,

[3]Menon, V.P., *The Story of the Integration of the Indian States*, Orient Longman, Chennai, 1956.

who, till the day he died, thirsted for the Valley. Mahajan remembers that Jinnah made three unsuccessful visits to Kashmir, ostensibly for health reasons, including sending his military secretary, an Englishman.[4] All three times, Maharaja Hari Singh had rebuffed him. Jinnah decided to turn to other methods to coerce the Maharaja into acceding to Pakistan.

The Maharaja of Kashmir signed on the dotted line in Jammu and formalized the structure of the accession of his state to India after packing his bags to flee to the safer haven in the state. Meanwhile, Jinnah, who treasured Kashmir as the big prize of his two-nation theory, tried his best to reach out to Sheikh Abdullah one last time, knowing fully well that even Abdullah did not like him, but the tall Kashmiri leader was his best bet to poach Kashmir.

As the raiders knocked on Srinagar's doors, the real drama shifted to New Delhi. At the behest of the British Home Office Political Department, Jinnah made several attempts to get in touch with Sheikh Abdullah at 17, York Road, which was Pandit Nehru's residence then.

Abdullah arrived in Delhi to appeal to the Indian Cabinet to dispatch armed forces to help Kashmiris repel the Afridi tribesmen. Jinnah wanted Abdullah to visit Lahore for a dialogue. But alas, he left it too late. It was late October 1947, and the Indian Cabinet was in an emergency session, considering whether to send troops to Kashmir or not. Sardar Patel recommended strong action, and he was supported by Defence Minister Sardar Baldev Singh.

---

[4]Mahajan, Mehr Chand, *Looking Back: The Autobiography of Mehr Chand Mahajan*, Asia Publishing House, Delhi, 1963.

Sheikh Abdullah, sitting in the anteroom with K.N. Bamzai, officer on special duty (OSD) to Nehru, slipped a piece of paper to an angry Nehru saying that the Indian Army needed to be sent quickly or all would be lost. Things calmed down, the Defence Committee met again and green-lighted the Indian assault on Srinagar. Abdullah was to fly with the first sortie of 40 Dakotas on the morning of 27 October, where the National Conference workers ferried them in buses from the tarmac.

At the same time that decisions in Delhi were being made regarding the future of Kashmir, the locals, the National Conference leaders and the armed forces of the state of J&K came together to form the Red Cap Militia to repulse the invaders. Tales of bravery from those turbulent times live to this day. As the Kabali and Afridi raiders hurtled down the Muzzafarabad Road to converge on Srinagar, at Uri, Brigadier Rajinder Singh, who led J&K State Forces, was martyred after a long, pitched battle against the odds. It was a scene out of Greek mythology where the brave Leonidas and his troop of 300 soldiers held the Persian invaders at bay at Thermopylae. Similarly, Maqbool Mohammed Sherwani, a martyr, patriot and a National Conference member, became a symbol of this exemplary defence for delaying the raiders at Baramulla. He misguided the rebels, and for this, was brutally murdered. These battles and tales of gallantry were vital, for it helped Maharaja Hari Singh avoid capture and bought the anxious Indian government valuable time to fly in their forces.

A young colonel by the name of Sam Manekshaw was privy to all that was happening in Delhi. He recounts the tension-filled meetings in Delhi, with V.P. Menon and

Mehr Chand Mahajan being ferried to and fro between Srinagar and Delhi, and the heated meetings between Nehru, Abdullah and Patel regarding the best way forward. According to Manekshaw, the armed forces were ready to depart for Kashmir. But, while they had been ready for a few days, they could only be flown in once Kashmir had acceded to India. For the time being, Manekshaw was sent to survey and take stock of the situation before reporting back to Mountbatten and the Defence Committee. He hurried back with the news that the raiders were about 7 or 8 miles away from Srinagar, and if troops were not flown in immediately, Kashmir could be lost.

> There were tense times, it was all touch and go. The raiders were upon Srinagar, Mahura Power station had been burnt, the Maharaja had fled in the dead of the night, Nehru was vacillating, Sheikh was pressing him, Army was on standby. According to Manekshaw, V.P. Menon was in the meantime discussing with Mahajan and the Maharaja. Eventually the Maharaja signed the accession papers and we flew back in the Dakota late at night. There were no night facilities, and the people who were helping us to fly back, to light the airfield, were Sheikh Abdullah, Kasimsahib, Sadiqsahib, Bakshi Ghulam Mohammed, D.P. Dhar with pine torches, and we flew back to Delhi. I can't remember the exact time. It must have been 3 a.m. or 4 a.m.[5]

---

[5]Panthaki, Behram M., and Zenobia Panthaki, *Field Marshal Sam Manekshaw: The Man and His Times*, Niyogi Books, Delhi, 2014.

As soon as the Instrument of Accession was ratified, Nehru and Patel gave the orders for the troops to be flown into Srinagar. The Sikh Regiment under the command of Lieutenant Colonel Dewan Ranjit Rai was the first to land, followed by several more columns of soldiers as the battle for Kashmir began. It was a close call as the airfield in Srinagar was in jeopardy of being overrun by the invaders, which would have made troop mobilization impossible.

The problem emanated from what noted journalist H.N. Brailsford aptly described years ago:

> The peculiarity of Kashmir's case is that the ruler and the subjects are of different creeds. The mass of the population, 93 per cent of it (in the Valley), is Muslim, whereas the Hindu Maharaja governs through a ruling class of Dogras and Rajputs drawn from his ancestral domains of Jammu. This was at the core of the issue for no Kashmiri may enter his army and there is an Arms Act which permits Rajputs to own firearms, a privilege denied to men of other stocks. The lower ranks of its civil administration were until recently the monopoly of Kashmiri Brahmins, though latterly, room has been made for Muslim candidates.[6]

If that was at the core of a complex accession, throw in Jinnah's manic obsession to poach the 'Eastern Switzerland' and Pandit Nehru, as a Kashmiri himself, loathe to giving it away, and you have all the ingredients of a high-octane drama. If that wasn't all, then add two dynamic

[6]Brailsford, H.N., *Subject India*, Gollancz, London, 1943.

protagonists—Sheikh Mohammad Abdullah and his bête noire, Maharaja Hari Singh, and you have a bitter tale of a power tussle fuelled by a Muslim-dominated and long-subjugated Muslim-majority Valley opposed to a Hindu-ideology-backed Jammu province.

1

SHEIKH MOHAMMAD ABDULLAH: THE SHER-I-KASHMIR

No story of modern Kashmir can be told without the tale of the tallest political personality—Sheikh Mohammad Abdullah—who battled the monarchy to bring its subjugated people out of serfdom. In Kashmir's pantheon, he is undoubtedly the dominant force, and no narrative on Kashmir can ignore him. As he and others like him—Hakim Ali, Safdar Hamdani, Mirza Afzal Beg, G.M. Sadiq and others—returned from Aligarh and other educational centres, they were attracted to progressive ideas and a quest for a democratic system to form a middle-class intelligentsia with a middle-class political outlook. Their early and tentative meetings became known as the 'Reading Room' discourses, since the forming of political associations was banned in the state. These reading groups later assumed a political dimension but were initially set up to improve the educational facilities in the state and promote a scientific temper. G.A. Ashai set up the first Islamia Old Boys

Association (a reading room) with 20 members, including Abdullah, who was still a student, in 1922.

Isolated among the snow-capped mountains and enveloped in its own garb of social and political backwardness, Kashmir at that time was as backward as most Indian states. People—and what abject pictures of mass suffering they were—had learnt to labour for the benefit of their soul and for the good of the masters, who seemed destined to rule. And so, the tradition of habitual servitude passed on from father to son, and none questioned the tyranny and exploitation that seemed a part of the social system of those days. The upper classes and the privileged communities lived to enjoy and exploit, and their tradition of serving the government as its instruments passed down from generation to generation.

Everything looked to be predestined, and nature herself seemed to favour this steady process of exploitation. Winter was a long period of virtual hibernation when all active outdoor life came to standstill. With spring and summer, life returned again, and Kashmir once more put on a festive appearance. It all meant some trade, work in the fields, plenty of marriages and a good rush of visitors. Occasionally, some preachers from the plains of Punjab came and delivered unintelligible discourses or some unusual event in British Indian politics excited a short-lived curiosity among the few intellectuals. A notable example of this is the Punjab-based group Majlis-e-Ahrar-ul-Islam, which started visiting the state regularly. Within Punjab and northern India, this political group had been extremely active and involved with the Khilafat Movement of the early 1920s. But once in

Kashmir, the annual visitors knew and learnt nothing beyond what concerned them. They came and enjoyed and went back to their homes. Occasionally, they wrote about Kashmir and its natural charm—the Jhelum's crystal-clear water and shimmering lakes; its snow-covered peaks and gorgeous landscapes; its glaciers and waterfalls. A few with a scholastic turn of mind indulged in learned discourses about the ancient history of Kashmir—about the age of Kalhana and the justice of Zain-ul-Abidin. But none of them felt that underneath an apparent calm, Kashmir was a mighty volcano—a vast land of seething discontent and human suffering.

That was Kashmir before 1930. In the summer of that year, Sheikh Mohammad Abdullah, a young man of hardly 25 and a fresh graduate from the Aligarh Muslim University, returned to his home in a half-neglected suburb of Srinagar. Abdullah had graduated from Islamia College, Lahore, and then went on to pursue a MSc degree at the Aligarh Muslim University. On his return to the Valley, he found a job as a school teacher, and together with a few more young men, decided to dedicate his whole life to the cause of the masses. Close to 100 years of Dogra rule had significantly impoverished the common populace and civil liberties, since the ascension of Pratap Singh, had almost disappeared. A mighty volcano of mass unrest was soon to burst open, and no power on earth could have stopped it then. The climax had been reached, and it only needed an efficient leadership to give it an organized expression. The movement, thus, had a gigantic urge behind it, and it gave point and direction to the thoughts and aspirations of young Sheikh Abdullah and his followers.

When Maharaja Hari Singh returned from overseas, a public meeting took place in Khanaq-i-Mohallah, an important sacred site in Srinagar, on 25 June 1931 to ratify the Muslim representatives who were expected to submit grievances and the demands of the community at the behest of G.E.C. Wakefield, political member of the Imperial Council. This became a defining moment in the crusade against the oppression of monarchy, as Sheikh himself articulated that this was the formal inauguration of the freedom movement of Kashmir. The Muslim representatives were ratified here through a show of hands. Cries of 'Allahu Akbar' rent the air as Sheikh took oath. It is here that Sher and Bakra—Abdullah and Moulvi Muhammad Yusuf Shah—both spoke with great fervour to free the Muslim majority of serfdom. Subsequently, Sheikh and Moulvi went their separate ways. Sheikh, a famed orator, showed his skill and asked the Muslims and Pandits to join the cause—'The present state of affairs has caused me and my colleagues a great deal of hard thinking and appraisal of the situation which is in no way satisfactory.'[1]

A mass movement, thus, developed, and its details have since become a part of the fabric of history. Following the public proclamations in 1931, the All Jammu and Kashmir Muslim Conference was formally established in October 1932. A Jammu-based Hindu ruler dominating a Muslim-majority Valley was anathema to the masses, due to the Maharaja's prejudices and domineering attitude. A

[1]Khan, Nyla Ali, *Sheikh Mohammed Abdullah's Reflections on Kashmir*, Springer, Delhi, 2018.

gargantuan mass movement, inevitably communal in the beginning but popular and nationalistic during the latter phase of its development, shook Kashmir from end to end and challenged age-old conceptions and habits. It revolutionized the entire outlook of the people and brought about a complete change in their social and political thinking. It presented a new and progressive aspect of life and made people conscious of their rights and privileges. Kashmir, which, over the centuries, had been ruled by the Sultan dynasty, Mughals, Afghans, Sikhs and Dogras in succession, was readying to overthrow the yoke of the monarchy.

More than anything else, the movement sought to unite the disparate and diverse communities of Kashmir. These were, thus, in brief, the achievements of the Kashmir movement of which Sheikh Abdullah became the architect and the nurturer. A most conspicuous and unique person, Abdullah, more than anyone else, merited the adoration of the masses. His was a dedicated life of suffering and sacrifice, and the masses followed him blindly and worshipped—there is no better word—him like a demigod. Tall, with a strapping personality and excellent rhetorical ability, Abdullah engaged with the masses and became their great leader. Academic and author Nyla Ali Khan writes:

> Popular mobilization in Jammu and Kashmir during the 1930s and 1940s took the form of uprisings, which was a primary locus of political action. This potent political resistance was led by people like Sheikh Mohammad Abdullah, Chaudhry Ghulam Abbas, Mirza Afzal Beg, Maulana Masoodi, and Kasap Bandhu, who did

> not have access to the echelons of power and spoke vociferously from the margins. Their activism made substantive forays into established discourses and structures of power.[2]

By 1945, the National Conference movement achieved an inflection point, as Abdullah and his team created a wide network of regional and rural committees, with peasants and the middle class also joining in. Large-scale demonstrations took place with two dominant slogans: 'Tear up the Amritsar Deed Sale of Kashmir'[3] and 'Quit Kashmir'. Ramesh Chandra Agrawal, writing on the events, concluded, 'Kashmir is the centre of the fight for independence. The imperialist policy followed by Britain for safeguarding her power in India, now rests on the support of the Princes. You can notice that in the plan of the Cabinet Mission, in the new India–British investments, and in the creation of the armies and aerodrome in the principalities.'[4]

At the court of the sessions judge, Srinagar, in the fabled 'State versus Sheikh Abdullah' case, where he was put on trial under Section 124A Indian Penal Code, Sheikh Abdullah, charged with sedition, spoke fiercely in his own defence. The speech was a landmark one in the battle against monarchical rule:

---

[2]Ibid.

[3]The Treaty of Amritsar, signed between the British East India Company and Raja Gulab Singh of Jammu after the first Anglo-Sikh War on 16 March 1846 established the suzerainty of the British Empire in the J&K State.

[4]Agrawal, Ramesh Chandra, *Kashmir and Its Monumental Glory*, Aryan Books, Delhi, 1998.

> I have pleaded not guilty to the charge of 'sedition', which according to jurists, is a crime against society, and I stand by whatever I have said or written in regard to the fundamental rights of the people of Jammu and Kashmir. The reports of my speeches, though neither verbatim nor strictly accurate, are fairly correct. But when reliance is placed in the complaint and the charge, on certain sentences, torn out of their context and not free from distortion, a wrong impression is likely to be produced. I owe it to myself and the decencies of civilized controversy to state that I have maintained a clear and strict distinction between persons and policies, between individuals and the system of government which they operate, and I would not allow my undignified or indecent reference to or vulgar abuse of anyone anywhere. I have, however, all along sought the alteration of policies and measures and a vital change in the system of government by legitimate and civilized means, for it is the birthright of man to shape and mould the law by which he must live.[5]

He further added, in what was a typical Sheikh Abdullah tour de force where he tore into Maharaja Hari Singh's regime, one that had subjugated the vast populace of the Valley:

> I am not interested in personal defence, and I would not have undertaken it if I had not felt that my trial for

[5]Kaul, Santosh, et al. (eds), 'Document 102', *Kashmir: Legal Documents*, Light & Life Publishers, New Delhi, 1977.

> 'sedition' is something far more than a personal charge against me. It is, in effect, a trial of the entire population of Jammu and Kashmir, even though some of them, being content with their transient personal interests or out of fear, may not be prepared to recognise or openly declare this. Oppressed by the extreme poverty, misery and lack of freedom and opportunity of the people of Jammu and Kashmir State, I and my colleagues of the Jammu and Kashmir National Conference, many of whom are behind prison bars or in exile today, have humbly sought to serve them during the past sixteen years. We have endeavoured to give faithful expression to the growing consciousness among the people of their imprescriptible rights, aspirations and desire for freedom.

Sheikh Abdullah's sources of influence and power were many, and he never tried to exploit them for his personal benefit. Consider first the heritage which he derived as the leader of the mass movement that changed the very shape of Kashmir. Its social and political advantages were immense, and it left a permanent impact on the lives of the people. Secondly, consider his attitude to the vast political gains that accrued or would accrue as a result of the political awakening. While his party accepted office, and he became the prime minister of the state, he still found sufficient time to live among the masses as their trusted friend and guide. Thirdly, his personal life became a source of great inspiration to his followers and the people at large.

A staunch Muslim himself, Abdullah took equal delight and pride in the religion and the religious festivities of other communities. During these festivities and events, he made sure to attend public gatherings convened on the occasion and delivered learned discourses on the teachings of Hindu and Sikh gods and prophets, which were listened to with great interest and fervour by everyone present. His was an unbiased and a robust patriotism, and his love of human liberties and sentiment made him a stout patriot and a humanitarian. Lastly, imagine the appeal that his simple and unostentatious life must have had for the people. He lived an austere life and neither drank nor harboured any vices. His dress was spartan and his way of life was exemplary.

Sheikh Abdullah was the single dominant factor in the political movement of Kashmir. It would, therefore, be no exaggeration to any that he became the National Conference itself, subsuming the party with his stature. The mighty task of converting the vast and well-organized communal movement into a national force and broad basing it on the people as a whole was accomplished by him, and it will remain a marvel of political ingenuity and great imagination in the political history of India. While it constituted a great milestone in the political development of Indian states, it contributed greatly to the strengthening of popular and progressive forces in the north and northwest of India. No rival Muslim communal organization found it possible to start its work in Kashmir ever since the Muslim Conference changed its name to the National Conference under Abdullah's guidance.

Several attempts were made by influential quarters, to which may also be added the last serious attempt of

Mohammed Ali Jinnah to start a Muslim communal organization in Kashmir. But the pull and influence of Sheikh Abdullah and the National Conference scuppered any such plans. Jinnah came to Kashmir as the self-appointed guardian of the political rights of the Muslims of Kashmir and mistook Abdullah's hospitality and courtesy for his acceptance of the supremacy of the Muslim League. But he was mistaken and had, in consequence, to face a very embarrassing situation. As Abdullah boldly declared that Jinnah was a nobody in Kashmir politics, the masses of Kashmir expressed their indignation and disgust with Jinnah's tactics by making it impossible for him to address their meetings. Finally, a despondent Jinnah beat a hasty retreat, and Abdullah's National Conference won over the forces of reaction, theology and narrow communalism.

The secret of Sheikh Abdullah's power was his vast influence with the masses and his deep understanding of the complexity and peculiarity of Kashmir, which had escaped Jinnah's formulation regarding the state. This understanding helped Abdullah free his fellow countrymen from their narrow fanaticism and other prejudices. His popularity has been admitted by everybody who visited Kashmir during the period. In the villages of Kashmir, legends came to be woven round his personality, and it was difficult to remove the impression from the minds of the people that Sher-i-Kashmir was not endowed with some supernatural powers. Whenever he was on tour, vast multitudes would travel long distances to see him and perhaps touch him, just to be sure that Abdullah was real and not a dream.

Outside Kashmir, Sheikh Abdullah was a towering personality of national importance. For Nehru, Abdullah was the magnet to bring Kashmir into India's fold. A Kashmiri himself, Nehru wanted to showcase his secularism model by using Kashmir as a shop window. For that, Abdullah was the man.

At the same time, in the war against monarchism, Sheikh Abdullah was used as a catalyst to expedite Praja Mandal reforms in the states, something that the rulers were loathe to doing. The people's organizations started by the people of princely states against colonial rule were called Praja Mandals or Praja Parishads or Lok Parishads, and they fought the feudal rajas and maharajas to kickstart the process of democratization. The Hitvardhak Sabha was founded in Poona in May 1921, while Wamanrao Patwardhan founded the Akhil Bhor Sansthan Praja Sabha in Bhor in November 1921. Nehru's All India States Peoples' Conference began in 1927, with its first session in Bombay in 1927, bringing representatives from hundreds of princely states. Nehru smartly appropriated Abdullah and turned him into the showpiece of the All India States Peoples' Conference, a stalking horse created by him to make inroads into the princely states by taking the Congress message deeper and deeper into these cloistered regions. Though Abdullah mostly championed the cause of the princely states' people in his avatar as vice president of the All India States People's Conference, he always showed a keen interest in Indian politics. While he was not a Congress member organizationally, he was a staunch Congressman both politically and ideologically. He was a firm believer in the unity and freedom of India, and while he admitted and even celebrated the diversity of social and cultural interests of the different people, he stoutly stood against all attempts aimed at breaking the political unity of India as a whole. Thus, he opposed Jinnah's two-nation theory vehemently.

If Kashmir acceded to India, it was primarily due to the patriotic mindset of Sheikh Mohammad Abdullah, who adopted an antagonistic approach against Jinnah's sectarian and communal thinking. Both were tall Muslim leaders, both were alumnus of the Aligarh Muslim University but were diametrically divergent in their worldviews.

2

ALL THE KING'S MEN

Had he acceded to Pakistan before August 14, the future Government of India had allowed me to give His Highness an assurance that no objection whatsoever would be raised by them. Had His Highness acceded to India by August 14, Pakistan did not then exist and therefore could not have interfered. The only trouble that could have been raised was by non-accession to either side, and this was unfortunately the very course followed by the Maharaja.

—Lord Mountbatten speaking at the East India Association in London on his return from India

If Sheikh Abdullah and the National Conference were the protagonists of what eventually became a brutal cage match, then their opponents were the rulers and rule-makers of the kingdom of J&K—Maharaja Hari Singh and his cabal of courtiers.

The general perception was that Maharaja Hari Singh was a British puppet, as all the important posts in Kashmir were occupied by trustworthy men from the British Political Department. The resident was Colonel Webb, the commander-in-chief of Kashmir State Forces was Major General Henry Lawrence Scott, while the inspector general of police was also a Britisher, Richard Powell. The truth is somewhere in between. As a dynast, Hari Singh held no fondness for Jinnah and his Muslim League, who were intent on creating their own nation along communal lines. As a Hindu, the Maharaja knew that he wouldn't be looked upon favourably in this new country. On the other hand, he was deeply distrustful of the Congress and Nehru, in no small part due to their proximity with his bitter enemy, Sheikh Abdullah and the National Conference. Hari Singh did not equally trust the British, though he played along with them, since they held the reins of power and were responsible for creating the concept of a 'State Subject'.

On 9 July 1925, soon after being crowned as maharaja of Jammu and Kashmir, Hari Singh declared, 'My religion is justice.' This might have annoyed several religious groups, yet Hari Singh stood by his commitment. On 20 April 1927, he issued a royal decree introducing a great revolutionary legislation, the State Subject ordinance, which has since been renamed as 'Permanent Resident'. The ordinance not only granted people of Jammu and Kashmir class 1 citizenship, but in order to protect rights of his citizens, affirmed that 'in matters of grants of the State scholarships, State lands for agricultural and house building purposes and recruitment to State service, State subjects of Class 1 should receive preference over other classes...'[6]

While he held a more progressive worldview than his predecessors, as shown in his declarations and policies when he started, Hari Singh was still a king. And as a king, his priority after the first few years became the preservation of his rule, and along the way, he and his courtiers made several blunders that led to Kashmir still being undecided about its future on 15 August 1947.

One of the most prominent personages in the court of Hari Singh was Ram Chandra Kak, the PM of the princely state of J&K, and a faithful servant of the autocracy. He knew that in the wake of the Second World War, Anglo-Americans were the main spearhead of the new western axis, and they were putting heavy premium on every reactionary force in every corner of the world. The experience of Greece, China, Turkey and Iran had taught all reactionaries that they could count on an endless supply of dollars and war material if they were sufficiently ruthless in suppressing their own people. Kak's policy was to save the autocracy by selling out to imperialism. This could be done only suppressing the people's movement ruthlessly, something that he showed great appetite for.

Kak, who had been in positions of power in British departments before he became the PM in 1945, knew fully well why the British had reared 550 odd princes all these years. He knew why so much solicitude was being shown for the rights and privileges of these 'ancient houses'. He knew

[6]Kaul, Santosh, et al. (eds), 'Document 44', *Kashmir: Legal Documents*, Light & Life Publishers, New Delhi, 1977.

their job was to form a network of friendly fortresses in a foreign territory. He saw that Kashmir was to be one of the main bulwarks of this intricate chain, for Kashmir was in a pre-eminent position to play a key role in the geopolitics of the region, wedged as it was on top of India and between what would be Pakistan, and touching the borders of China and Afghanistan. Kak understood that in the new world order where the Anglo-Americans would oppose the Soviet Union bitterly, Kashmir could become the perfect base for anti-Soviet operations. Its key importance, according to Kak, was only enhanced by the Mountbatten Plan, which gave assent to the partition of the country into two new nations that would join the British Commonwealth.

This new dynamic situation opened up a plethora of possibilities for Kak. Inside the state, the national movement could be killed by the wave of communalism engendered by the boost it got from the 3 June Plan. One could play the two dominions against each other and possibly hold them at bay by deceiving both. The good offices of Lord Ismay, Sir George Cunningham and Sir Robert Francis Mudie were available for sowing the seeds of conspiracy, while Inspector General Powell, Major General Scott and Colonel Webb were vital to keep the liaison going to ensure everything that was being done was off the record. All these office-bearers were complicit with Kak. National Conference leaders were securely locked up once Kak declared martial law in response to the National Conference's Quit Kashmir movement in 1946. With imperialism covering your back, who could dare challenge autocracy? This was the frame of mind in which Kak met Acharya J.B. Kriplani, who was visiting Kashmir to convince the Maharaja to accede to India in May 1947.

Major General Shahid Hamid of the Pakistan Army, in his account of Partition, describes Kak in so many words:

> Kak, a Kashmiri Brahmin himself took over the prime ministership from Sir Gopalswamy Ayyanar ICS. He wanted to introduce reforms in the State, but Hari Singh resisted. Thereafter they fell apart and there was no love lost between them. On no occasion was his advice accepted. Kak...was a man of strong character, completely western in outlook. He spoke fluent English, a man of the world, he was married to an English woman. Nehru thought him too clever and hated him. Kak was a realist and openly advocated that Kashmir should join Pakistan. He maintained that the State would get a fair deal and this would save it from the turmoil of partition.[7]

There was an uncanny cogency in Kak's chain of thought, and he showed some efficiency in its execution too. But he ignored one crucial imponderable altogether—he was totally blindsided by the organized strength of the people's movement both inside and outside the state. By simply not factoring in the people's upsurge in his calculations or the width and depth of the crisis in the imperialist camp, he missed a trick or two. According to Hamid, Kak continued warning Hari Singh that the people of Kashmir

[7]Hamid, Shahid, *Disastrous Twilight: A Personal Record of the Partition of India*, Leo Cooper, London, 1993.

would rise against him if he ceded to India. Colonel Webb also maintained that it was logical for Kashmir to accede to Pakistan.[8] Kak, in fact, warned Hari Singh that there was danger of Muslim tribesman descending on the Valley from the mountains, which is exactly what happened finally. The outcome was that Hari Singh locked himself in the palace and ignored Kak's advice of carrying out negotiations privately. At the same time, he started cashing out his assets in Kashmir and transferring them to India and the United Kingdom, and he started a dialogue with Vickers, a British engineering company, to buy an amphibious Viking aircraft for his safe passage out of the state.

Kak sensed that the ground beneath his feet was slipping away. He saw imperialism incapacitated for open intervention with men and material to keep its puppets in power. He believed that the British would not leave their watchdogs at the mercy of the wolves, so he indulged in a lot of table thumping. Soon, the cold truth dawned on him when he saw the Nizam pleading with the British, 'I cannot believe that, after more than a century of faithful alliance, the British Government, it is the intention of the British Government to throw my State out of the Empire against my will.'[9]

If Ram Chandra Kak was seen as the Maharaja's closest adviser, even though Shahid Hamid claims otherwise, there was another power behind the throne, which many considered far more potent than Kak. In a lot of ways,

---

[8]Ibid.

[9]Poplai, Sundar Lal, *India, 1947-50: Issued Under the Auspices of the Indian Council of World Affairs*, Oxford University Press, 1959.

this person was seen as one of the major reasons behind Hari Singh's indecision regarding his kingdom. He was the rajguru, Swami Sant Deo, a man who regularly stoked the Maharaja's feudal ambitions. Muhammad Yusuf Saraf, former chief justice of Azad Kashmir High Court, in his controversial book *Kashmiris Fight for Freedom*, writes:

> He was not a citizen of Jammu and Kashmir State and probably came from Punjab. Shapely in appearance, he first attracted attention in Srinagar during the last years of Maharaja Pratap Singh. He used to dress himself in costly silk and given to extensive use of perfumes... He spent a few years in a Gurukul in Kangrah with the Arya Samajists who conferred on him the title of Swami. He met Pratap Singh on one of his visits to Hardwar and this is how he got into Kashmir. He took up residence in a government bungalow near Chasma Shahi, though for some time he also resided at the Mughal-built Bara-Darri in Achaal.[10]

Saraf further writes that as an initially progressive ruler, Hari Singh, unlike his predecessor Pratap Singh, allowed no room for superstition and promptly dismissed Deo as soon as he took the throne. The fanaticism that characterized Pratap Singh's rule was gone, replaced by able-minded courtiers and intellectuals, with whom Hari Singh wanted to bring about major reform in Kashmir.

However, with the years, contends Saraf, insecurity and

---

[10]Saraf, Muhammad Yusuf, *Kashmiris Fight for Freedom: 1819–1946*, Volume 1, Ferozsons, Lahore, 1977.

the preservation of the monarchy became more important to Hari Singh, and with the rise of the National Conference, he retreated more into his palace. It was at this time that his wife, Maharani Tara Devi, started wielding an influence upon her husband. During the 1940s, according to Saraf, Hari Singh was entirely dependent upon the Maharani. Tara Devi was a devout Hindu, and keen to foster spiritual and religious initiatives, she would often invite members of Hindu organizations to the palace. She soon, apparently, became an ardent disciple of Deo and his ideas. In this atmosphere, Swami Sant Deo returned to the kingdom and retook his position as rajguru and started influencing matters of state policy, something he had no expertise in. Karan Singh, in his *Autobiography*, recounts:

> It was in the political sphere that Swamiji's influence proved to be disastrous. As with many of the larger States, the prospect of becoming an independent ruler after the British withdrawal was an alluring one for my father. He had never taken happily to British suzerainty, and at the same time was far too deeply enmeshed in the feudal tradition to really come to terms with democratic forces that were gaining momentum in the subcontinent and within the State itself. A major weakness of the feudal system is that the Rulers are told only what their courtiers think they want to hear, which is seldom in consonance with what is in fact happening. It was on this feudal ambition that Swamiji astutely played, planting in my father's mind visions of an extended kingdom sweeping down to Lahore itself, where our ancestor Maharaja Gulab Singh and

> his brothers Raja Dhyan Singh and Raja Suchet Singh had played such a crucial role a century earlier.[11]

Singh further states that as 1947 wore on and the monarchy came under attack from several quarters, not the least of which was the pressure of choosing a dominion to accede to, the Maharaja felt disillusioned with his rajguru and subsequently saw and listened to him less and less.

Another one of Hari Singh's courtiers who had also been appointed on the auspices of Maharani Tara Devi was the deputy PM, Ram Lal Batra, who, in Karan Singh's estimation, was not up to the task. Saraf, on the other hand, states that Batra was a fanatical Hindu, and the reason why Tara Devi chose him over an Englishman, who was the previous minister-in-waiting to Kak, was because both the Minister and the Maharani were in favour of acceding to India. This was an odd mixture that had the ingredients to cause an implosion: The Maharani and Batra were all for joining the Indian dominion, whereas Kak, the man in power, was in favour of acceding to the dominion of Pakistan. Hari Singh, on the other hand, was being influenced by Deo to try holding on to an independent kingdom.

In the days leading up to the accension of Kashmir to India, there was constant coming and going between Srinagar and Delhi by trusted emissaries. Some ruling princes from East Punjab and Congress leaders visited Kashmir to persuade the Maharaja to join India. Meanwhile, Hari Singh also

[11]Singh, Karan, *Autobiography*, Oxford University Press, Delhi, 2003.

started negotiations with the Sikh mountain states on the eastern border to allow him to build a road that would bypass Gurdaspur and give him direct access to Delhi. These ideas showed a schism in the Maharaja's mind that would come back to bite the entire kingdom later on. He was unable to decide, even as late as early 1947, whether to accede to one country or to remain independent.

A significant section among the Kashmiri masses believed that the policy of Congress leadership with regard to Kashmir had serious consequences (with the exception of Nehru, who was emotionally driven) and did not do much to drive confidence in the political as well as the royal leadership of J&K. Congress president Acharya Kripalani's visit to Kashmir in May 1947 is considered a glaring example. Instead of supporting the just demands of the Kashmiris, Kripalani had tepidly asked them to stop their slogan and protest of Quit Kashmir and to acquire self-government in a peaceful way under the patronage of the Maharaja. This confused policy of the Congress incited the undecided Maharaja to think in terms of remaining independent and continuing with his repression of the popular movement led by the National Conference.

On 18 June 1947, Lord Mountbatten paid a visit to Kashmir to persuade the Maharaja to not make any declaration of independence but to opt either for Pakistan or India latest by 14 August 1947. The Maharaja conveyed to Lord Mountbatten that he did not want to accede to either but wished to be independent. Lord Mountbatten returned disappointed, and, ultimately, Nehru decided to send Mahatma Gandhi for reaching a deal with the vacillating Maharaja.

On 1 August 1947, Gandhi arrived in Srinagar for a meeting with Hari Singh and Tara Devi. He gave them full assurance that in case he acceded to India, his interests as head of the state would be protected. Further, he advised him to release Sheikh Abdullah and also remove Ram Chandra Kak from prime ministership, as he was against the accession with India. The next day, Gandhiji visited Soura to meet Begum Abdullah, where he assured her that Sheikh Abdullah would be released soon.

In his letter to Nehru dated 6 August 1947, Mahatma Gandhi gave some details of his talks with Maharaja Hari Singh, Maharani Tara Devi and Ram Chandra Kak. He wrote:

> During my interviews with the Prime Minister, Ram Chandra Kak, I told him about his unpopularity among the people. He wrote to the Maharaja that a sign from him he would gladly resign. The Maharaja had sent me a message that he and the Maharani were anxious to see me. I met them. Both admitted that with the lapse of British Paramountcy, the true paramountcy of the people of Kashmir would commence. However much they might wish to join the Union, they would have to make the choice in accordance with the wishes of the people. How that could be determined was not discussed at the interview...[12]

Gandhi's visit proved to be of great success to Nehru's plans. Though Maharaja Hari Singh gave up the idea of

[12] 'Gandhi Ji in Kashmir', Daily Excelsior, https://tinyurl.com/5n79p5pm, Accessed on 12 October 2022.

an independent Kashmir, Gandhi's visit could not bridge the gap between Hari Singh and Nehru. The Maharaja was apprehensive that in case he acceded to India, Nehru would take revenge of his arrest in Kashmir. He also feared that Nehru would hand over power to Sheikh Abdullah, who was the Maharaja's sworn enemy. Accession with India would also mean Pakistan would block all supplies, and state would be in great trouble. The Indian Union would not accept accession unless the National Conference was satisfied, for Nehru, Sheikh and the National Conference were vital to showcase a secularist India viz a theocratic Pakistan, which owed its creation to religion. Perhaps Kak's dismissal would satisfy the National Conference and make them waive their claims to higher stakes. Perhaps the sham of bitter anti-Kakism would create the necessary atmosphere in which the National Conference leadership could be persuaded to be satisfied with a few crumbs of power. This was the autocracy's game plan. Ram Chandra Kak was asked to step down on 11 August, and a politically weak General Janak Singh was appointed as the new PM.

As the autocracy recast its strategy with Janak Singh and Batra taking over from Kak, with the former representing the outlook of the Dogra autocracy and the latter the game plan of the RSS, the common folk of Kashmir thought it was a huge victory and awaited Sheikh Abdullah's release from jail, which did not happen until 29 September 1947.

In this backdrop, the Maharaja announced his intention to sign a standstill agreement on 12 August 1947 both with Pakistan and India. Telegrams were sent to India and Pakistan by PM Janak Singh. Pakistan came to the rescue by recognizing the Maharaja's sovereign status and signing the standstill agreement unconditionally, it even offered to guarantee the privileges of autocracy. But openly acceding to Pakistan was out of the question, as the National Conference had still not been completely crushed and the loyalty of the Dogra Army was dubious.

The new strategic imperative of the autocracy was to escape the transfer of power to the people by weakening the national movement from the inside. It gambled on the growth of communalism within the state and used the inter-dominion play as an escape valve for itself. The scheming of the autocracy created doubts and misgivings even about the bonafides of the National Conference leadership. Not even Sheikh Abdullah was spared these mischievous insinuations. Although he had publicly declared that he had not entered into negotiations with the government during his imprisonment, doubts were cast on the veracity of his statements. All this heartened the autocracy, as the situation was fluid, and by clever manoeuvring, they succeeded in creating an impression of a secret deal, a plan began unfolding like clockwork. But Sheikh Abdullah, darling of the masses, kept the attention of the people riveted on the basic and fundamental issue, namely, of ending the autocracy. He refused to be drawn into the controversy surrounding the issue of accession. Without mincing words, he condemned the two-nation theory. At the same time, he drew a clear distinction between acceding to Pakistan and an ideological surrender to Jinnahism. The question of accession, he rightly left to the sovereign will of the people, who would decide according to what they wanted.

Nehru was Sheikh's alter ego in this battle for supremacy against Hari Singh. His views and public declarations were important in mobilizing public opinion against the monarchy not only in the mainland but also in the kingdom itself, which put Hari Singh's court at the centre of some unwanted attention. Take his views on 27 May 1946 as a sampler:

> Kashmir affords a notable illustration of how things must not be done. So long as there was some attempt at cooperation between the State administration and the National Conference, unsatisfactory as it was, there was no possibility of trouble. The moment this policy was reversed on the part of the State, the situation underwent a progressive change for the worse...As soon as Kak became PM 11 months ago, one noticed the beginnings of a new policy, for the new PM was deliberately preparing to fight and crush ruthlessly and without apology the Kashmir NC. The Maharaja, who kept out of local politics has given a blank cheque to the present PM and others, extends itself now to the Maharaja also for he is ultimately responsible.[13]

Battle lines had been drawn, and sides had been picked. The three participants were the royalists in favour of independence, the National Conference in alliance with the Indian dominion, and the Muslim Conference in alliance with the dominion of Pakistan. The stage had been set for an explosive and fierce battle that would break apart the state of J&K forever.

---

[13]Sharma, Suresh K., *Encyclopaedia of Kashmir*, Anmol Publications Pvt Limited, 1996.

3

THE TWO DOMINIONS

A confidential note prepared by Jawaharlal Nehru's private secretary, Dwarkanath Kachru, dated 8 June 1947 provides an insight into Kashmir's integration, which stands to this day:

> Geography, it has been suggested in certain influential quarters, will no doubt play a great part in determining the course which states would follow. Due to the proposed division of Punjab, it is therefore permissible to argue that Kashmir will find itself geographically isolated from the rest of India. Powerful influences may also develop which will tend to draw Kashmir to the new alignment of forces within areas adjacent to it. But considering the temperament and character of the people of the state and the policy and outlook of the National Conference, it can safely be assumed that under the stress of disintegrating forces, Kashmir would tend to develop progressive nationalism and a

> political life of its own rather than join any unhappy combination in the areas adjacent to it.[14]

Kachru's words were so prescient. They stand the test of time, and even 75 years later, reverberate through the corridors of every Kashmir expert's memory recesses. The idea and ideal of Kashmir were integral to both Nehru and Sheikh Abdullah, and that remains at the very kernel of Kashmir's accession to the Union of India.

Attempting to scupper this was Mohammed Ali Jinnah, who arrived in Kashmir to a hero's welcome in May 1944. His visit in response to an invite sent by the Muslim Conference leaders saw him spending close to two months in the Valley. Shunned by the ruler and his PM, he was welcomed by Abdullah, who was famously part of a picture with Jinnah celebrating at Aligarh Muslim University; however, it did not stop Jinnah from making incendiary speeches. Take a look at one of them and this highlights his obsession with Kashmir, 'If your objective is one, then your voice will also become one. I am a Muslim and all my sympathies are for the Muslim cause. It is better to live in a hut in Pakistan with a sense of security than to live in a bungalow in India in a shadow of security.'[15] Finally, in his address to the Muslim Conference session held in the compound of Jama Masjid on 17 June 1944, he said, 'It is my duty as a Muslim to advise you correctly as to which course would be proper

---

[14]Documents from the private collection of K.N. Bamzai.

[15]Kaul, Santosh, et al. (eds), 'Document 83', *Kashmir Legal Documents*, Light & Life Publishers, New Delhi, 1977.

and ensure your success.'[16] For close to two months, he canvassed across the length and breadth of the Valley, but all to no avail.

After the plan to partition the country and create the two dominions of India and Pakistan was announced on 3 June 1947, Jinnah resumed his machinations, making every effort to cause a rift between Maharaja Hari Singh and the Indian leaders, attempting to prevent the state's proposed accession to the Indian Union. To facilitate this process, he sent a detailed set of instructions to the Muslim Conference in Kashmir that they should advocate sovereign independence for the state and oppose tooth and nail any suggestion made by other political parties in favour of accession to India.

The Muslim Conference leaders who, in the normal course, were anxious that the state should accede to Pakistan could not follow the significance of this directive. So, Chaudhary Hameedullah Khan, president of the Muslim Conference, and a few other leaders met Jinnah in Delhi. Jinnah's plan was simple—as Kashmir claimed independence and refused to accede to the Indian Union, it would be quite easy to force its accession to Pakistan at an appropriate time. They were then told to give all possible assurances to the Maharaja that the Muslim Conference would stand by the state's independent status so much so that Hameedullah Khan made a public statement that if ever Pakistan tried to invade Kashmir, he would be the first person to take up arms and fight Pakistan.

---

[16]Ibid.

After the declaration of the 3 June Plan, not just Kashmir but most of the 550 princely states were in a similar conundrum about whether to listen to the political parties that had cropped up in the last few decades and accede to one or the other dominion, which the populace demanded, or to hold on to their sovereignty in any way possible and find a compromise there. While the British had announced their intention to transfer power, political officers brought up in the old imperialist traditions were thinking in terms of a Third Bloc or Third Force that would play their game, pinning their hopes on the princely order for that end. As the Cabinet Mission announced that paramountcy would lapse on the transfer of power to Indian hands, the issue that faced all princes was whether to join the Constituent Assembly or stay independent. In 1946, there were two distinct groups among the princes—firstly, those who planned to join the Constituent Assembly, and the others who remained undecided. The Nawab of Bhopal was the spearhead amongst the undecided. Jinnah and the Muslim League took advantage of this position to sow the seeds of further discord amongst the princes and convince them to ally themselves with Pakistan and maintain their thrones at the same time.

On the 3 June Plan, like the Nawab of Bhopal, Ram Chandra Kak was heartbroken, saying privately that the British had let the princes down. Kak subsequently increased his engagement with Jinnah and the Nawab of Bhopal. And his plan for Kashmir's independence hinged on support from Jinnah and the League. Even Mountbatten's visit to Hari Singh did not yield any result, for Hari Singh continued to hold out for the possibility of a separate entity independent of both Pakistan and India. Jinnah assured Kak through the Nawab of Bhopal that not only would Pakistan respect Kashmir's sovereignty but Pakistan would also help in every possible manner if India refused to sign a standstill agreement with the state and imposed economic sanctions. The foreign minister of Bhopal, Mr Qureshi, was the intermediary who visited Kashmir many times to convey these secret assurances. When Mahatma Gandhi was slated to visit Kashmir to meet and convince the Maharaja to accede to the Indian Union, the Muslim Conference general secretary asked for Jinnah's instructions, whether there should be hostile demonstrations or not, and in reply, Jinnah sent a cryptic message, 'Leave that rogue to the tender mercies of Kak.'

Fortunately, evidence of the Bhopal foreign minister's secret missions and also some letters written by Jinnah to Kak fell into the Maharaja's hands, and it resulted in Kak's sudden dismissal (pressure from Gandhi during his visit to Srinagar was another impetus) just before 15 August. Jinnah, shrewdly understanding that his instructions were miscarried, sent Sheikh Sadiq Hassan and Nawabzada Rashid Ali Khan, Muslim League leaders from Punjab, to Kashmir with instructions that the Muslim Conference should reverse its stand and launch a campaign demanding the state's immediate accession to Pakistan.

Nicolas Mansergh, writing in his authoritative *Survey of British Commonwealth Affairs,* says:

> Kashmir lay at the frontiers of both successor States, it was neighbour to both, and while its ruler was Hindu,

> its people were predominantly Muslim. There was no State in the subcontinent the future of which, if left unsettled when the independent imperial authority withdrew, was more likely to occasion dissent between the two Dominions. The failure to concentrate more closely upon the problem it presented was destined to prove an oversight fraught with grave consequences.[17]

Positions were taken, and soon more players entered the game, attempting to influence opinions and communalize the already sensitive issue. The prominent Pakistani newspaper *Dawn* wrote on 30 July 1947, attempting to build opinion, 'The only sensible course for Kashmir is to join the Pakistan Dominion in which it can retain far greater freedom that it can as integral part of the other Dominion.'[18] Supplementing this was the communiqué of 11 August 1947 which said, 'The Government of Pakistan recognizes Kalat as an independent sovereign State in treaty relations with the British Government with a status different from that of the Indian State.'[19]

In his book *The Last Days of the British Raj*, Leonard Mosley confirms that Maharaja Hari Singh operated with a single-mindedness of purpose—not to align with either

[17]Mansergh, Nicolas, *Survey of British Commonwealth Affairs: Problems of Wartime Cooperation and Post-War Change 1939–1952*, Routledge, London, 1968.

[18]Quoted in Krishen, Rajbans, *Kashmir and the Conspiracy against Peace*, People's Publishing House, Delhi 1951.

[19]Veena, Yogeena, 'How Balochistan Became a Part of Pakistan—A Historical Perspective', *The Nation*, 5 December 2015, https://tinyurl.com/2p82f52m. Accessed on 13 October 2022.

India or Pakistan and seriously consider independence as an option. So, he followed a policy of keeping his mouth shut and letting Kak and, later, other PMs of the state do the front running and, for most part, do the talking. Moseley writes:

> Pandit Nehru had no reason to love the Maharajah, who stood for everything which the Indian leader hated; he was a bigot, a profligate and an autocrat with small concern for the people of his State; and he had shown his contempt for the Congress by suppressing its members, jailing its leaders, and threatening arrest to Nehru, himself a Kashmiri Brahmin, if he dared set foot inside the State. Yet Nehru sent messages to the Maharajah advising him not to decide too precipitately on what he should do. Gandhi emerged from his preoccupation in the South and sent a similar message.[20]

As detailed earlier, both J.B. Kriplani and Gandhi did go to Kashmir to meet the dodgy Maharaja. Mountbatten insisted that he would go too, since he was an old friend of Hari Singh. They had both been ADCs (aides-de-camp) on the staff of the Prince of Wales when he toured India in 1931. On 21 June 1947, Mountbatten arrived in Srinagar and stayed at the palace. He took George Abell with him, and in the next 48 hours, for a negotiator who could be a typhoon of energy when he wished, he accomplished extraordinarily little. His press secretary and biographer Alan Campbell-Johnson notes:

---

[20]Mosley, Leonard, *The Last Days of the British Raj,* Weidenfeld and Nicolson, 1961.

> When he got there he found the Maharaja politically very elusive and the only conversations which took place were during their various car drives together. Mountbatten on these occasions urged him and his Prime Minister, Pandit Kak, not to make any declaration of independence, but to find out in one way or another the will of the people of Kashmir as soon as possible, and to announce their intention by 14th August, to send representatives accordingly to one Constituent Assembly or the other. He told them that the newly created States Department of India was prepared to give an assurance that if Kashmir went to Pakistan this would not be regarded as an unfriendly act by the Government of India. He went onto stress the dangerous situation in which Kashmir would find itself if it lacked the support of one of the two Dominions by the date of the transfer of power.[21]

Campbell-Johnson went on to say that it was the Viceroy's intention to 'give this advice privately to the Maharaja alone.' It was decided that such a meeting would take place on the last day of the visit. 'But when the time came the Maharaja sent a message he was in bed with colic and would be unable to attend the meeting. It seems that is his usual illness when he wishes to avoid difficult discussions. Needless to say, Mountbatten is very disappointed with these turn of events.'[22] What was remarkable about this episode

---

[21]Campbell-Johnson, Alan, *Mission with Mountbatten*, Macmillan, London, 1985.

[22]Ibid.

was Mountbatten's fatalistic acceptance of the Maharaja's evasions and subterfuge. One could understand if Nehru or Gandhi had been rebuffed in this manner and accepted his prevarication, they had everything to gain by it; after all, time was still on their side. And the moment Sheikh Abdullah would be released by Hari Singh in panic or under duress, the game would change instantaneously in Kashmir, for the tall nationalist leader was a votary of accession.

So, why did Mountbatten, who bulldozed his way with other princes, including the chancellor of the Chamber, the nawab of Bhopal, vacillate on Kashmir? Here was a state whose future was more likely than any other state to be contentious, since it was wedged between India and Pakistan, and equally Tibet, China, Afghanistan and Russia. Geopolitically, a decision on Kashmir was paramount, and how could Mountbatten miss the primacy of such a decision? More so, Mountbatten, a soldier himself, understood the import of India holding on to Kashmir, since its frontiers had been breached repeatedly in the past by invaders. Why did his flair desert him on this all-important occasion? Mosley asks, 'Can it be that, exhausted by his constant round of talks, sickened by Jinnah's evasions over the problem of Governor Generalship of Pakistan, allergic for the moment to the Muslims, he was suffering from a diplomatic colic too?'[23]

Another version goes like this. Alex von Tunzelmann, writing in *Indian Summer,* notes:

---

[23]Mosley, Leonard, *The Last Days of the British Raj,* Weidenfeld and Nicolson, 1961.

> On June 18, the Mountbattens flew to the beautiful lakeside city of Srinagar, for a visit to the largest of the princely states, Kashmir. Nehru had already warned Mountbatten that the Muslim-majority Kashmir might prove to be a problem. Mountbatten knew the Hindu Maharaja, Sir Hari Singh, having first met him in Calcutta at Christmas dinner in 1921, for which Sir Hari had been specially selected as one of the British government's best behaved princely allies. Their acquaintance had been sufficiently cordial that Dickie and Edwina had holidayed with him in Srinagar as late as 1946...Mountbatten assured the Maharaja that he would be allowed a free choice of which dominion to join after 15 August, emphasizing the importance of ascertaining the will of the Kashmiri people, but making it clear that any moves to remain independent would be foolish and dangerous.[24]

Following Kak's dismissal and the appointment of Mehr Chand Mahajan as PM in the days before the declaration of independence, there was a feeling on both sides that the state of J&K would accede to India. Even though nothing had been given in writing, the Maharani, along with the presence of pro-India ministers Mahajan and Batra, made it seem apparent to both the Congress and the Muslim League that there was a high chance of the state joining the Indian dominion. This led to much unrest and frustration among the Pakistanis, which prompted Jinnah to devise one final

[24]Tunzelmann, Alex von, *Indian Summer: The Secret History of the End of an Empire*, Simon and Schuster, London, 2012.

nefarious plan to grab control of Kashmir.

As part of Jinnah's larger plan, the general public in Poonch and Mirpur districts, who were mostly ex-servicemen, were supplied with arms and ammunition. In a matter of a few weeks, the stage was set for widespread internal strife as soon as the signal was given. Attempts were made to start communal rioting in Srinagar on the pretext that the Pakistan flag was insulted, but these efforts didn't materialize. Further, all the important leaders of the Muslim Conference were asked to leave Kashmir and carry on their subversive activities from Pakistan so that they could not be arrested by the Kashmir State Forces. Muslim Conference leaders and some workers started leaving Kashmir in batches and assembling in Pakistan, from where they established contact with their party workers in the border districts. The ex-servicemen rose in revolt in early September, as if in synchronicity, and fought pitched battles with the state troops. Between June and October 1947, the Poonch rebellion took place, an uprising against Maharaja Hari Singh led by the Muslims, who were a martial race, which had started as a 'No Tax' campaign. This was followed by armed bands crossing over from adjoining Pakistan into the state to help the insurgents. A rigorous economic blockade of Kashmir from the Pakistani side was then set in motion, drying it off of all essential commodities.

Around the same time, the Pakistan government started secret negotiations with National Conference to woo it and win it over as Pakistan jump-started the final offensive to pouch the Valley. The DC of Rawalpindi and Dr Taseer, then an employee of the West Punjab government, were sent to Srinagar to contact Sheikh Abdullah, who remained aloof, to persuade him to support the state's accession to Pakistan. After preliminary discussions with these emissaries, Abdullah sent Ghulam Mohammed Sadiq to Rawalpindi for obtaining specific terms from Pakistan. Earlier, Bakshi Ghulam Mohammad, another top National Conference leader, had met Pakistani Prime Minister Liaquat Ali Khan, the provincial premiers of West Punjab and the North-West Frontier Province (NWFP), besides other senior officials. Liaquat Ali Khan is believed to have given an assurance that if Abdullah agreed to Kashmir's accession to Pakistan, he would head the provincial government in Pakistan.

The plan was as follows: Sheikh was to make a public statement that in view of the geographical location and communal composition of Kashmir, the state should accede to Pakistan. If the Maharaja refused to accept his advice, then Abdullah and his colleagues were to cross over to Pakistan, where they were to set up a provisional government of Free Kashmir in Pakistan, and under its auspices, a large-scale invasion of the state was to take place. In their hearts, the Pakistan managers for Kashmir knew that the plan would not work, for Sheikh abhorred Jinnah and loved Nehru. So, an alternative plan was put together. Sheikh would be invited to Karachi for personal discussions with Jinnah, and if he failed to come around, he was to be secretly arrested and kept in an unknown place, and a provisional government of Free Kashmir formed under his presidentship, while he would rot in jail.

Sheikh Abdullah had conceived a dream vision of Kashmir with a political dispensation in which, regarding matters concerning to Kashmiris, their voice would be supreme and they shall have complete ownership on their resources. This arrangement, he had worked out, to confederate J&K either with India or Pakistan, with complete internal autonomy, limiting the accession to three things—defence, foreign affairs and communication only. As a precondition, he wanted a constitutional guarantee to this effect.

In view of above, on 1 October 1947, a delegation from Pakistan headed by Dr Taseer had a meeting with Sheikh Abdullah, Mirza Afzal Beg and others at the Majestic Hotel, Lal Chowk, Srinagar, where Sheikh Abdullah, in categorical terms, told Dr Taseer, 'We would join Pakistan on the basis of Naya Kashmir programme and State's accession to Pakistan would be limited, i.e., Defence, Foreign Affairs and Communication only. Pakistan shall have to promise a guarantee to our internal autonomy.'[25] In the meeting, it was decided that G.M. Sadiq would accompany Dr Taseer to Pakistan for further discussions with Mohammed Ali Jinnah.

Nehru was very watchful of these developments, and he invited Sheikh Abdullah to Delhi. He agreed to all of Abdullah's conditions and assured him that his dream vision was possible in India only, as he knew his temperament well and never wanted to offend him at this critical point of time. Dr Taseer and Sadiq reached Pakistan on 3 October, where Bakshi Ghulam Mohammad was already waiting for them. A meeting was arranged with Pakistani leaders and also with Liaqat Ali Khan on 8 October 1947. Abdullah directed Maulana Muhammad

[25] Documents from the private collection of K.N. Bamzai.

Syed Masoodi, another National Conference leader, to leave for Lahore immediately to take part in the negotiations between Sadiq and the Muslim League government.

On 23 October, Sadiq was informed by Liaqat Ali Khan that Jinnah was not willing to meet Abdullah or negotiate any terms, and insisted that Kashmir should join Pakistan unconditionally. Sadiq insisted on a meeting with Jinnah. The meeting was yet to take place when Kashmir was invaded by tribal men of the NWFP of Pakistan. Sadiq's dialogue with Pakistan had now become irrelevant, so he left for Delhi to join Sheikh Abdullah. Sadiq informed Abdullah about the developments in Pakistan, who already had some misgivings about Pakistan's feudal landlords and was sure that they would create hurdles to implement his 'Naya Kashmir' manifesto. As Abdullah's mind and heart was with Nehru, these developments lead to throwing his weight in favour of India. As all this drama unfolded and Sheikh continued to play truant, the D Day for the invasion kept being postponed. Abdullah instead reached Delhi.

The blueprint for the invasion had been drawn up much earlier. Liaquat Ali Khan summoned a top-secret meeting in Murree in the beginning of September, which was attended, among others, by the nawab of Mamdot, Khan Abdul Qayyum Khan, Major Shaukat Hyat Khan, Sir Abdur Rahim, General Akbar Khan, Colonel Sher Ahmed Khan, Major General Shahid Hamid, Air Commodore M.K. Janjua, Mian Mohammad Abbas and a number of Indian National Army officers. Plans for the invasion were readied (even today the same modus operandi is being followed, Pakistan still uses the stratagem of pushing infiltrators to create confusion and ill will amongst Kashmiris in the Valley) and it was decided that ex-servicemen from Poonch numbering 80,000 veterans would form the bulk of the attacking force. The two provincial premiers of Punjab and the NWFP, the Nawab of Mamdot and Abdul Qayyum Khan, were asked to provide all the necessary assistance to these military officers to collect, train and transport the necessary personnel to various points for a lightning attack. Mian Mohammad Abbas (accountant general of Pakistan military) was asked to provide the necessary funding. Orders were issued to the army authorities to make available necessary arms, ammunition and other equipment required for such an invasion.

The two-pronged strategy was for Liaquat Ali Khan to conduct political negotiations with Sheikh Abdullah under his direct supervision, while the military officers were asked to speed up their arrangements for the invasion. Poonchis and other ex-servicemen were gathered in military training camps around Rawalpindi and given strenuous refresher training and held in readiness for the final attack, while in the meantime, small armed raiding parties attacked the state's border towns and villages, forcing a dispersal of the state forces. The progenitor of the military strategy was Sir George Cunningham, the governor of NWFP from 1937 to 1946 and again, at the invitation of the governor general of Pakistan, from 1947 to 1948 after a year in retirement; he was the go-to man in Pakistan. In *Mission with Mountbatten*, Campbell-Johnson explicitly articulates, 'Indian opinion has been inclined to suspect Cunningham of machiavellian designs and of secretly sponsoring the diversion of tribes into Kashmir.'[26] Sir George wanted Pakistan to make a dramatic move, which would stir up religious fervour of the tribesmen and strengthen the bonds of Islamic fraternity amongst the tribes of Pakistan. Since the situation in the tribal areas was fast deteriorating and Afghan propaganda to create a Pathanistan was on the rise, Sir George thought it would be better to channelize their energy into Kashmir.

To wean the Pathan tribesmen away from Afghan influence, he planned this alternative strategy. At a meeting in Rawalpindi, which was attended by Liaquat Ali Khan, Sir Francis Mudie, Nawab of Mamdot, Qayyum and Sir George, besides top military officers, such as General F.W. Meeservey and General D.D. Gracey, the NWFP governor laid bare his sinister plan. The main raiding force, it was decided, would be the tribesmen, and propaganda would be pumped up in the tribal areas that Kashmir had to be freed as part of a jehad (yes, it is still the same messaging). Qayyum and Cunningham suggested that the launch pad should be a 10-mile strip of land on the India–Pakistan border, which was evacuated by Punjabi villagers for fear of border raids from East Punjab, but with the West Punjab government not prepared to take up this responsibility, the plan was scrapped, and the Muzaffarabad road selected.

Philip Woodruff, ICS, writing in *The Men Who Ruled India*, provides a rather benign view of Cunningham:

---

[26]Campbell-Johnson, Alan, *Mission with Mountbatten*, Macmillan, London, 1985.

> He was Governor of the North-West Frontier Province from 1937 to 1946 and again, at the invitation of the Governor-General of Pakistan, from 1947–48. For the first three years of his time, he had a strange problem of a Muslim Congress ministry. Everything was topsy-turvy on the Frontier and when, in the twenties, people of the settled districts—not of course the tribesmen across the border—came to know that they might play a new game called politics, they began with enthusiasm. They were a healthy and highspirited people and their idea of politics and their idea of politics was to embarrass the Government. It was easy to form an anti-Government party; what was not so easy was to find a programme or grievances or, turning it into jargon, an ideology—and someone to quarrel with.[27]

Woodruff believed that there was really only one genuine political question in the administered territory of NWFP; the Khans or chiefs retained some feudal privileges, which were out of date, and they were still in a position to exercise a good deal of influence. It was possible to draw a kind of party line between the friends of the Khans and their opponents, and the new anti-government party took the side of the opponents. Sir John Maffey, an *amirparwa*, a protector of the nobles, had made a belated attempt in the early 1920s to increase the power of the Khans, and that had sharpened the feeling. And in the villages

[27]Woodruff, Philip, *The Men Who Ruled India: Volume 2: The Guardians*, Johnathan Cape Ltd., London, 1971.

where that distinction had no meaning at all, two factions who opposed each other for the mere sake of strife would choose one party or the other. In the 1920s, in that part of the world, the Muslim League was a decorous and law-abiding body; the opponents of the Khans and of the government, therefore, chose to affiliate themselves to the Indian National Congress, and in that improbable partnership, they remained until politics suddenly became serious in August 1947.

According to Woodruff, when the Muslim League came into office in 1937, they were led in the NWFP by Dr Khan Sahib, a man who had been surgeon to a regiment of the Indian Army and had not the least personal animosity either to British officers or to European ways. Like the other Congress premiers, most of his views were a good deal modified when he came into office, and it was seldom that any real controversy seemed likely with the Governor, who was on the whole the protector of the poor and felt sympathy with him over the Khans. When difficulty did arise, Sir George would ask him round to play bridge and lose a few rubbers amicably, after which it was usually possible to come to an agreement. Woodruff further notes:

> In 1947, when the new Dominion of Pakistan was set up, Sir George, who had been a year in retirement, was invited to come back as Governor, being the man everyone trusted. The time that followed he found the most interesting, and also the most peaceful, of his career. It was so much better without the Government of India; there was so little paper, so few instructions.

> Mr Jinnah, the Governor-General, wrote to him, he thought, once only.[28]

The reality was that the frontier tribes were armed, supplied with ammunition and even with trucks and fuel with the knowledge of the British commander-in-chief, General Masservy, and as many as 6,000 British officers were in charge of the Pakistan arsenals. As part of the conspiracy theory, it was believed that the military posts situated between the frontier territory and the border of Kashmir, which could have easily checked the raiders, were conveniently evacuated. The task of beginning the invasion was entrusted to Kuli Khan, a well-known British agent in the tribal regions. Three of the White masters of Kashmir—Webb, Scott and Powell—left the state not long before the invasion. More interestingly, the chief of the general staff of the Kashmir Armed Forces, Rajendra Singh, disappeared without any trace, while the administrative apparatus built so lovingly by the British Political Department practically ceased to function.

Tearing apart Campbell-Johnson's hypothesis about him, Sir George Cunningham, who maintained a personal diary written in longhand, said with great emphasis that he had tried to prevent the invasion at every possible stage. Cunningham was worried at 'quite a lot of talk of the actual danger of war.' On 6 October, he confided to himself:

> The Pakistan government seems to me to wink at very dangerous activities on the Kashmir border, allowing small parties of Muslims to infiltrate into

[28]Ibid.

> Kashmir from this side. That might easily become a casus belli...A week later, one can see the horizon darkening: There is a real move in Hazara for Jehad against Kashmir, I have warned everyone I could, including Afridis and Mohmands, of the danger of taking part in anything like this, in case it leads to war between India and Pakistan.[29]

So, either Sir George was Janus-faced or a consummate actor whose conscience was pricked after he had hatched the plot, one will never know. On 15 October, he wrote to Jinnah, 'I am perturbed by the move which a lot of our people are making in the Kashmir direction. The talk now is not merely of defence against attacks from Kashmir by Sikhs, but of invasion into Kashmir.'[30] On 18 October, he got his answer from NWFP strongman and chief minister, Abdul Qayyum, who put it bluntly that it would be a good thing to fill Kashmir with armed Muslims.

Preliminaries to the invasion of Kashmir and Jammu, which finally led the Maharaja and his people to accede to the Indian dominion, had begun soon after the partition on 15 August 1947. On 29 August, Maharaja Hari Singh received a telegram from one Raja Yakub Khan on behalf of the public of Hazara, alleging attacks on Muslims in Poonch and informing him that the Hazara Muslims were much perturbed. Throughout the month of September, reports of

---

[29]Khan, Nyla Ali, 'Women in Kashmir: The Glorious Roles They Played', *Epilogue*, Vol. 4, no. 3, 2010, p. 34, https://tinyurl.com/yeep6jes. Accessed on 13 October 2022.

[30]Ibid.

infiltration into the state territory from the border districts of Pakistan continued to come in. On 31 August, Major General Scott, general officer commanding of Kashmir State Forces, submitted a report regarding the situation in the state (he painted a grim picture). On 4 September, Scott reported that 500 hostile tribesmen wearing green and khaki uniforms with leather and web equipment were seen.

A protest was made directly to the Pakistan Army and to General O. de T. Lovett, the officer commanding the 7th Infantry division about gangs operating and raiding state territory. On 4 October, raiders armed with tommy guns were reported in Chirala, and many more were seen passing through the Jhelum. Similar activity was observed in Mirpur. Fort Owen was besieged and was evacuated by state troops on 15 October. On 18 October, the Kotli–Poonch road was breached and heavy fighting had begun. On 20 October, the Wazir of Mirpur sent a message that armed men were gathering opposite Chechiam and Kangla. Two days later, he reported that raids were being methodically carried out on Owen. On 23 October, heavy fighting was reported from Kotli, which, by now, had been completely cut off from Poonch by roadblocks put up by the raiders.

On 24 October 1947, the government of India received the first request for military aid from the government of J&K. At that time, the government had no agreement—military or political—with the state. The Kashmir War had begun.

4

THE KASHMIR WAR

The great state of J&K originally had five parts to it—Jammu, Kashmir, Ladakh, Aksai Chin and Northern Areas (Gilgit-Baltistan). Narendra Singh Sarila, writing in *The Shadow of the Great Game*, highlights how in 1935, the administrative and defence responsibilities of this northern frontier had been transferred by the Maharaja of Kashmir to the British Government of India under a 60-year lease.[31] As the result of the civil war in China became uncertain, the Viceroy had prevailed upon Hari Singh to do so in the interests of the security of the empire. The region of Gilgit was administered by the Political Department from Delhi in the same way as agencies in the NWFP, such as Malakand or Khyber, with political officers stationed there reporting to the Viceroy through Peshawar. A carefully chosen force capable of rapid movement in mountainous territory, controlled by British officers (called the Gilgit

[31]Singh Sarila, Narendra, *The Shadow of the Great Game: The Untold Story of India's Partition,* HarperCollins, Delhi, 2009.

Scouts) provided the muscle to the administration.

Fifteen days before Independence (on 1 August 1947), Gilgit was retroceded by Delhi to the Maharaja of J&K, and Lieutenant Colonel Roger Bacon, the British political agent, handed over the area to Brigadier Ghansara Singh, the newly appointed state's governor from Srinagar. According to V.P. Menon, 'the Kashmir authorities did not have the resources, including financial, to hold Gilgit which was cut off from Srinagar during winters.'[32] In view of the lapse of paramountcy, the retrocession was probably inevitable, but the fact remains that no sooner was Gilgit handed over to the Maharaja than it came under the mercy of Pakistan (through the NWFP).

The British officers of Gilgit Scouts, Major William Alexander Brown and Captain A.S. Mathieson, remained to serve the Maharaja as contract officers, though they continued to receive instructions from the political agents based in Khyber in Peshawar, which, after 14 August 1947, had become a part of Pakistan. While Singh, the new wazir or governor, occupied his official residence in the grandeur of importance, it was Brown and Mathieson who held the keys to power in Gilgit. Lieutenant Colonel Bacon, on transfer from Gilgit, was given the Khyber post. This ensured perfect coordination between the Gilgit Scouts and Peshawar. According to the *Bulletin of Military Historical Society of Great Britain*, the broad post-Partition plan had been discussed by Brown and Bacon in June 1947. And after

[32]Menon. V.P., *Integration of the Indian States*, Orient BlackSwan, Delhi, 2014.

Mathieson arrived in Gilgit as second-in-command, the two British officers refined contingency measures, should the Maharaja take his state over to India.

In such a situation, writes Sarila, whatever the fate of the rest of J&K, delivering Gilgit to Pakistan was fairly straightforward.[33] This was accomplished on the night of 31 October 1947, apparently according to an already-worked-out plan. As soon as Maharaja Hari Singh acceded to India, Major Brown got the Gilgit Scouts to surround the residency, and after a short gun battle in which he lost a scout, he imprisoned Governor Ghansara Singh. Peshawar was then informed by Brown about the accession of Gilgit to Pakistan. On 2 November, Brown raised the Pakistani flag at his headquarters and informed the force that they now served the government in Karachi. Brown and Mathieson had opted for service in Pakistan as the Maharaja signed the Instrument of Accession. Since Gilgit, by this act, had become a part of India properly, the officers should have made an immediate request for release from their appointments.

The actions of Brown and Mathieson were suspect politically, and while Brown describes it as a coup d'état, Alistair Lamb writes that Brown was certainly not acting as a party to a British conspiracy.[34] There existed, however, a small number of British soldiers and officials who, in their private capacity as friends of Pakistan, encouraged Brown

---

[33]Singh Sarila, Narendra, *The Shadow of the Great Game: The Untold Story of India's Partition*, HarperCollins, Delhi, 2009.

[34]Lamb, Alistair, *Kashmir: A Disputed Legacy (1846–1990)*, Roxford Books, London, 1991.

and Mathieson to be in Gilgit on the eve of the transfer of power. Moreover, what happened subsequently came as no surprise to someone like Lieutenant Colonel Bacon, who acted as a liaison between Major Brown in Gilgit and the Government of Pakistan, and in this respect, he may have contributed significantly to the success of the Gilgit coup. Bacon, however, in no way represented the policy of the British government in London. Neither Bacon nor, indeed, Colonel Iskander Mirza, defence secretary of Pakistan, were particularly unhappy when they heard about what was going on.

A document signed by the British chiefs of staff of the Indian armed forces states that on 24 October, information of the capture of Muzaffarabad was received by the Supreme Commander of India and Pakistan, Field Marshal Sir Claude Auchinleck. No plans for sending troops to Kashmir had, up to that time, been considered by the Indian Army. On 25 October, Government of India directed the preparations of plans for sending troops to Kashmir by road and air. Indian troops were sent to Kashmir on 27 October, following the signing of the Instrument of Accession. The accession was legally made by Maharaja Hari Singh on the advice of Sheikh Abdullah, leader of the Jammu and Kashmir National Conference, the political party commanding the widest popular support in the state.

Fearing for his kingdom and his life, Maharaja Hari Singh finally wrote to Lord Mountbatten on 26 October 1947:

> I have to inform your Excellency that a grave emergency has arisen in my State and request immediate assistance of your Government.

As your Excellency is aware the State of Jammu and Kashmir has not acceded to either the Dominion of India or to Pakistan. Geographically my State is contiguous to both the Dominions. It has vital economic and cultural links with both of them. Besides my State has a common boundary with the Soviet Republic and China. In their external relations the Dominion of India and Pakistan cannot ignore this fact.

I wanted to take time to decide to which Dominion I should accede, whether it is not in the best interest of both the Dominions and my State to stand independent, of course with friendly and cordial relations with both.

I accordingly approached the Dominions of India and Pakistan to enter into a standstill agreement with my State. The Pakistan Government accepted this arrangement. The Dominion of India desired further discussion with representatives of my Government. I could not arrange this is view of the developments indicated below. In fact the Pakistan Government under the standstill agreement are operating Post and Telegraph system inside the State.

Though we have got a standstill agreement with the Pakistan Government, that Government permitted steady and increasing strangulation of supplies like good, salt and petrol to my State.

Afridis, soldiers in plain clothes, and desperadoes, with modern weapons, have been allowed to infilter into the State at first in Poonch area, then in Sialkot and finally in mass in the area adjoining Hazara district on the Ramkote side. The result has been that the limited number of troops at the disposal of the State had to be dispersed and thus had to face the enemy at several points simultaneously that is has become difficult to stop the wanton destruction of life and property and looting. The Mahoora Power House which supplies the electric current to the whole of Srinagar has been burnt. The number of women who have been kidnapped and raped makes my heart bleed. The wild forces thus let loose on the State are marching on with the aim of capturing Srinagar, the Summer Capital of my Government, as a first step to over running the whole State.

The mass infiltration of tribesmen drawn from the distant areas of the NWF Province coming regularly in Motor Trucks using Mansehra-Muzaffarbad road and fully armed with up-to-date weapons cannot possibly be done without the knowledge of the Provincial Government of the NWF Province and the Government of Pakistan. In spite of repeated appeals made by my Government no attempt has been made to check these raiders or stop them from coming to my State. In fact both the Pakistan Radio and Press have reported these occurrences. The Pakistan Radio even put out a story that a Provincial Government has been set up in Kashmir. The people of my State both the Muslims and non-Muslims generally have taken no part at all.

With the conditions obtaining at present in my State and the great emergency of the situation as it exists I have no option but to ask for help from the

> Indian Dominion. Naturally they cannot send the help asked for by me without my State acceding to the Dominion of India. I have accordingly decided to do so and I attach the Instrument of Accession for acceptance by your Government. The other alternative is to leave my State and my people to freebooters. On this basis no civilised government can exist or be maintained. This alternative I will never allow to happen so long as I am the Ruler of the State and I have life to defend my country.
>
> I may also inform Your Excellency's Government that it is my intention at once to set up an Interim Government and ask Sheikh Abdulla to carry the responsibilities in this emergency with my Prime Minister.
>
> If my State has to be saved immediate assistance must be available at Srinagar. Mr. Menon is fully aware of the situation and he will explain to you if further explanation is needed.[35]

Even at this stage, when hostilities had begun between the two sides, the three British chiefs of staff of Indian forces warned of the risks involved in sending Indian troops. Nehru responded angrily that the only alternative to sending the troops would be to allow the massacre and loot in Srinagar, which would definitely be followed by an equally ferocious

---

[35]'The Maharaja of Kashmir's letter to the Governor-General of India, Lord Mountbatten, on October 26, 1947 and the Governor-General's letter in reply dated October 27, 1947', *Dawn*, 24 March 2012, https://tinyurl.com/knx3ak7y. Accessed on 13 October 2022.

response in other parts of India. Besides, the raiders would show no mercy to the hundreds of British residents in the state of J&K. So, the chiefs were silenced.

On 27 October, Mountbatten wrote back to Hari Singh:

> In the special circumstances mentioned by Your Highness, my Government have decided to accept the accession of Kashmir State to the Dominion of India. Consistently with their policy that, in the case of any State where the issue of accession has been the subject of dispute, the question of accession should be decided in accordance with the wishes of the people of the State, it is my Government's wish that, as soon as law and order have been restored in Kashmir and her soil cleared of the invader, the question of the State's accession should be settled by a reference to the people.
>
> Meanwhile, in response to Your Highness's appeal for military aid, action has been taken today to send troops of the Indian Army to Kashmir to help your own forces to defend your territory and to protect the lives, property and honour of your people.
>
> My Government and I note with satisfaction that your Highness has decided to invite Sheikh Abdulla to form an Interim Government to work with your Prime Minister.[36]

It was touch and go. The same evening, a message was sent to the Indian Army headquarters for airlifting some

[36]Ibid

forces to Srinagar. The first inkling of impending operation was received by the staff at the military command on the night of 26 October at 10.00 p.m. Never in the history of warfare had such an airlift operation been undertaken with so little notice and with even less planning. At that time, only four Dakotas were available with the air force. But the Indian Army and Air Force rose to the challenge to defend the country. About a hundred civilian and air force planes were mobilized during the night. The Royal Indian Air Force and civilian pilots, and ground crew worked heroically at night to launch the operation. Several pilots undertook many sorties during the day to make it a success. The civilian airline companies went out of the way to help the army in airlifting men, material and ammunition to stop the invasion. Six civilian Dakota transports took off from Safdarjung Airport. At the same time, three Dakotas of the Royal Indian Air Force were getting airborne from Palam airfield. By 10.00 a.m. on 27 October, the first elements of the Indian Army landed in Srinagar.

Major Khurshid Anwar, the 42-year-old Naib-Salar-i-Ala of the Muslim League National Guard, was leading the frontier tribesmen attack on Srinagar. Giving details of his campaign to *Dawn,* he had said, 'The attack on Kashmir was originally planned from two side-one from the regions adjoining Kashmir State in the tribal belt and the other from the Pakistan border.'[37] He claimed that the frontier tribesmen attack materialized but the other did not due

[37] *Documents on Kashmir Problem: White Papers, Domestic Jurisdiction and Resolutions,* Discovery Publishing House, 1991.

to obstructions placed by the Pakistan government. Many telegrams were exchanged between Kashmir's PM and Pakistani officials, where the details of the raiders' attacks were shared, but all to no avail. Sheikh Abdullah remained determined. He was heard saying at various forums:

> Kashmir is in dire peril, and the first duty of every Kashmiri is to defend his motherland against the intruders...the invasion of Kashmir is meant to coerce and compel the people of Kashmir to act in a particular way, namely, to acceded to Pakistan...every Kashmiri resents this compulsion on his will... Thirty lakhs of Muslims in Jammu and Kashmir State are not going to sacrifice themselves for one Nawab of Bhopal or one Nizam of Hyderabad whose interests the Muslim League is trying to guard.[38]

Underestimating Kashmiri resistance and realizing that the quarry may have got away from him, an impatient Jinnah sent his private secretary, Khurshid Ahmad, to Srinagar, but once the Indian forces arrived, he was arrested and sent back to Pakistan. V.P. Menon writes that as soon as Jinnah heard that India had accepted the accession of J&K and Indian troops had been airlifted to Srinagar, he ordered General Gracey, the acting commander-in-chief of the Pakistan Army, to rush troops to Kashmir for a full-fledged conflagration.[39] Gracey confessed to Jinnah his inability to

---

[38]Koul, Santosh, et al. (eds), 'Document 123', *Kashmir Legal Documents*, Light & Life Publishers, New Delhi, 1977.

[39]Menon, V.P., *Integration of the Indian States*, Orient BlackSwan, New Delhi, 2014.

issue any instructions in this respect without the approval of Field Marshal Auchinleck. He flew to Lahore on the morning of 28 October and explained to Jinnah that in the event of Pakistan troops entering Kashmir, which was now legally part of India, every British officer serving in the Pakistan Army would automatically and immediately be withdrawn. Jinnah then sent a message through Auchinleck to Mountbatten and Nehru, inviting them to Lahore for a conference to discuss the Kashmir problem. But Sardar Patel was strongly opposed to either of them making the visit.

Once the go-ahead came from the Indian Defence Council and planes airlifted Indian soldiers, the battle for the liberation of the frontier *illaqa*s (localities) of Kashmir began on 1 November. It was fought and won in the most spectacular manner and in the shortest possible time, with a remarkably negligible loss of life. It was fought amongst great hardship and handicap in tough terrain and in conditions that could perhaps be regarded as semi-arctic. The engagements took place in the deep narrow valleys of Gumri and Machoi, and in the hard 6-mile-long defile of the Batakundi feature, where the enemy was deeply dug in.

On 1 November, as the snow fell, and bitter frost and chill enveloped the valleys across the Zoji La, the first offensive was developed by the combined efforts of tanks and infantry. Examples of individual acts of heroism and reckless courage in this battle were many. In the highly tense situation before the final knockout blow while the enemy was firing from every cliff and every gorge, it was the general officer commanding, General Thimayya, who personally took charge and led the assault. He was at the head of the advancing columns and personally visited every section and brought the various columns—armour, infantry and gunners—into proper formation with a view to unleash maximum damage against the enemy. A brilliant tactician, he realized the necessity for firepower to blast the enemy out at the Zoji La heights, failing which the operation would be unsuccessful. Astonishingly, he ordered the deployment of a squadron of Stuart MK-VI tanks of the 7 Light Cavalry, compelling the engineers to cut a path through the rocky terrain in less than three weeks to build an almost 5-mile-long track that could take the width and weight of the tanks. Owing to the precipitous climb, the tanks had to be manually dismantled and pushed up by the jawans at an altitude where tanks had never been used before. General Thimayya himself had a narrow escape, and the death-defying tempo and mood of the occasion was enhanced all the more by his presence, compelling his men to dash forward regardless of the consequences.

The very appearance of the tanks utterly dampened the morale of the bewildered Pakistanis, who either fled or were swiftly vanquished. Once Zoji La was taken, the road to Kargil was reopened, and Ladakh was safe. After the bitter fighting in which the tanks played a prominent role, Indian troops entered Matayan on 3 November. The next battle, and which perhaps could be regarded as fierce as the previous one, lasted for a full 10 days. It was the grim battle for the Batakundi feature, and in this the men of the Royal Gorkha Regiment under Colonel Sukhdev Pathania and the Patialas under Colonel Sukhdev Singh played a notable part. The Gorkhas, in the freezing conditions of Matayan, waded through and back in the half-frozen stream and finally captured the important feature on this range, which was subsequently given the name of their commander—Anand Feature.

In this battle also, the tanks provided covering fire for the assault. After the fall of Batakundi, the enemy was at sixes and sevens as it ran following a rout. The Patialas, who were to march ahead after the capture of the feature and hold the LoC, gave them no rest. They were joined by other battalions, and soon Major Qureshi, the commanding officer on the enemy side, limped back in retreat after getting shot in the foot.

Indian troops then kept up the pursuit till they reached Kharal Bridge, 5 miles from Kargil. The enemy had crossed over to the other side and burnt the bridge down, thereby destroying its base planks. Moving on to higher ground, they took to sniping at advancing Indian positions. A little distance before Kharal Bridge, one of the platoons of the Royal Gorkhas led by Major Kochar crossed the half-frozen river in the dead of the night and sneaked up on the enemy, while another platoon of the Gorkhas marched ahead on the main road to Kargil. Taking advantage of the higher ground, the enemy killed five Gorkhas and wounded a few others. But that couldn't dampen the spirit or morale of the Indian troops, who kept pushing on. By now, the Pakistani army formations were in full retreat, and by 24 November, the Indian troops had cleared all the mountain ranges on both sides of Kargil Valley. Ladakh had been secured again. The same game plan was reprised in 1999 when they sneaked up and took high ground in the same area.

The enemy in this battle was not just out-manoeuvred and outwitted but was caught unawares by the presence of tanks and armoured columns. Pakistani officers found it difficult to believe that tanks could reach the Zoji La and kept on telling their men that the tanks were jeeps camouflaged as such. Thinking that they would not be evicted, they had stockpiled themselves with supplies, foodstuffs, clothing and uniform in Pandras, Drass and Kargil. Local gossip had it that they were ultimately left with only two boxes of ammunition and, therefore, had to withdraw and run away in disorder. But the truth of the matter was that they were shaken by General Thimayya's tactics and use of tanks at such high altitude.

This was a decisive blow in the ongoing war—the longest and most attritional war fought by the Indian Army. The defeat in the Kargil Sector naturally resulted in the complete withdrawal of enemy forces from the Leh Sector. The entire Zanskar and Nubra valleys were cleared by the Indian troops and militia. The civilian population everywhere between Zoji La and Kargil welcomed the liberators. Harrowing tales of extortion and mail administration by the enemy were narrated. The conquest of Kargil synchronized with the capture of Mendhar in the Jammu province. At this juncture, Pakistan thought India was going full tilt towards Gilgit in Kashmir and Kotli and Bhimber in Jammu.

The geopolitically sensitive Gilgit had been swallowed whole by two Brits acting in concert with Pakistan. Sir George Cunningham, the new governor of NWFP (whose role has been disputed in the sending of the tribal raiders), on hearing of Brown's coup in Gilgit, instructed him and his colleague Mathieson to restore order. Cunningham totally chose to ignore the fact that Gilgit was part of the J&K State, which had acceded to India.

Soon, a Major Aslam Khan, once deputy to Major Khurshid Anwar (one of the key Pakistan Army strategists who organized the tribal lashkar raid on Kashmir), arrived to take control over Gilgit. Throughout the Kashmir War, right from 22 October 1947 to 1 January 1949, when a ceasefire was proclaimed, which left Gilgit in Pakistan's hands, Britain successfully ensured that Pakistan's occupation of this region was not disturbed. After Mountbatten's mediatory role and the collapse of the direct talks between Nehru and Liaquat Ali Khan, the Indian Cabinet planned a full-scale war. But Mountbatten, who served India well all along, made a monumental blunder by suggesting to Nehru that the United Nations (UN) would promptly direct Pakistan to withdraw the raiders, which would make war unnecessary. And Nehru believed him, thus internationalizing the Kashmir issue. In the process, Gilgit was lost.

The Gilgit-Baltistan area or the erstwhile Northern Areas of J&K occupied by Pakistan covers 85,793 sq. km. It was further divided in 1970 into two separate administrative divisions: Mirpur-Muzaffarabad (which Pakistan calls Azad Jammu and Kashmir or AJK) and the federally administered Gilgit-Baltistan. Pakistan illegally ceded the Shaksgam Valley, around 5,180 sq. km, to China in a 1963 border agreement. China has been on a road-building spree in this area trying to connect the Karakoram Highway (built by it) with its ambitious Silk Road programme, which, in turn, will connect Xinjiang to Gwadar Port in Balochistan through the highway, a possible railroad, and an oil and gas pipeline. Strategically and geopolitically, the loss of Gilgit-Baltistan through the coup remains a major loss for India. In its quest to reclaim Kashmir, it erred in not going beyond Ladakh and capturing Gilgit-Skardu.

In an assessment from these areas, Nehru's private secretary and troubleshooter Dwarkanath Kachru wrote to the PM on 4 December 1948 saying,

> I don't know what reasons we have had for issuing instructions to halt our advance at this point and after discussions with Sheikh Abdullah, Bakshi and senior Army officers, I feel convinced that it would not be advisable to withdraw or even halt at the places which we are holding at present (don't know whether it is under UN pressure). I am personally convinced that the recapture of Kargil has committed us to the capture of Skardu because in order to hold the frontier illaqas of Ladakh and Kargil and to give the necessary security to the civil administration and the people, it would be necessary for us to liquidate the enemy pockets between Skardu and Ladakh and between Skardu and Kargil.
>
> In the Jammu province holding on to Mendhar alone would not be enough unless we try to go a little ahead and secure strategic and vital points which will ultimately give us the necessary DEGREE OF SECURITY. I WOULD HAVE ACTUALLY SUGGESTED A FULL BLAST ADVANCE TOWARDS MUZZAFARABAD because looking at the whole scenario from the approaching ceasefire, which seems

> to be inevitable, it would very advantageous to us if we have already reached and consolidated along the points and lines which we will ultimately demand as points and lines of demarcation between what we should have as our own and what may be left to be administered on our behalf.[40]

Calamitously, this advice was ignored.

As part of this continuing interface with Nehru, Kachru kept sending him confidential impressions from the Valley providing status reports. Earlier, on 8 July 1948, he had sent another note. While the focus of the note was on political content and the state of the Sheikh Abdullah administration, which was worse for the wear, it is the military side that was equally important, mainly regarding the morale of the troops:

> The military operations are going on successfully and according to plan and gradually large areas are being liberated and reclaimed. If, however, the request of General Thimmaya is acceded to and some more troops are loaned to him, we may in the very near future find ourselves in the occupation of Muzzafarabad and a substantial part of Poonch Jagir.
>
> Our troops have fortunately for us created a very good impression on the whole and are generally popular with the civilian population. But there is a peculiar feeling gradually creeping in the minds of our soldiers and the officers and whenever I had an

[40]Documents from the private collection of K.N. Bamzai.

> occasion to visit them at the fronts, they have spoken to me freely and frankly hoping that their feeling would be communicated to the Government. They feel that our Government is not taking the matter of Kashmir very seriously and are not fighting the war as it should be.'[41]

It would be safe to assume that PM Nehru was fully apprised personally of the military situation on the ground in Kashmir. It is also clear that things were going well, and Indian troops were poised to strike deeper and further in reclaiming parts lost in the immediate aftermath of the tribal incursion, as they evicted squatters successfully. But Nehru appeared hesitant to move forward more aggressively militarily despite repeated reminders and exhortations from Kachru (incidentally, also a Kashmiri), given he was advised by Mountbatten to take up the matter with the UN.

The dispatches from the war zone were frequent. In yet another note, dated 17 June 1948, when Kachru visited Tragbal along with General Thimayya and Deputy CM Bakshi, he notes:

> The General in his short exposition recorded in Tragbal informed those gathered that he had swept the whole of Handwara area including the whole of Karna Tehsil which included the recently liberated areas of Titwal and Karen. He hoped to liberate the entire area of every raider by the end of this month and then launch his offensive towards Gurez. At the foot of Tragbal rested

[41]Ibid.

> the Wular lake and it was one of the most beautiful spots in Kashmir. From Targbal, it is 16 miles to Gurez, the immediate objective of our Army. We have actually gone to a point over seven miles from Tragbal and the remaining 15 or 16 miles to Gurez may, we hope, be conquered in a couple of weeks.
>
> Gurez is an important town of very great importance both from the strategic and political viewpoint. Gurez is 155 miles from Gilgit and its recapture will open the road to Gilgit to our troops. Gurez at present is also the main supply base for the raiders operating in the frontier illaqas of Kargil, Ladakh and Skardu and is connected by a direct diagonal route to Drass, another important stop on the Srinagar–Ladakh road. Gurez in the hands of the raiders is therefore a great menace to us and its recapture will deal a death blow to the designs of Pakistan in the Sind Valley. The capture of Gurez should synchronise with the capture of Drass for then alone the Pakistani hordes will be trapped between these two points.[42]

Earlier, on 13 April 1948, another missive was sent to PM Nehru. Again, it showed how the battle was raging, and the complications regarding logistics and communication.

> With the onset of summer, military operations on both sides are assuming importance. In the Jammu province, the fall of Jhangar a few weeks ago, followed

---

[42]Ibid.

by a vigorous drive towards the adjoining area and particularly towards the town of Rajouri captured by the raiders in November last. Our troops entered Rajouri yesterday afternoon and the raiders ran away leaving the town looted, burnt and pillaged. The rejoicing of the people of Rajouri at the sight of Indian Army was reminiscent of the celebrations which took place in Baramulla when it was captured by us.

The recapture of Rajouri has a great significance as it relieves the pressure on Shopian, a town 35 miles south of Srinagar and on Ramban, a small station on the Jammu–Srinagar road. It also cuts off one of the supply lines of the raiders operating in the areas of Mirpur and Poonch. In Kashmir Valley, the battle on the Uri front has again flared up into a large-scale engagement and for the last one-week fighting is going on in various pockets in Mahura and Uri.

The situation on the Skardu front is, however, somewhat confusing and has to be tackled immediately. The Skardu garrison is holding on to the Fort in the town but it is practically cut off from the areas surrounding it. The pickets which were sent out are mostly missing and one or two of them have with great difficulty been able to make occasional contact with the Fort. The relieving columns sent from Srinagar have not reached yet and are still five stations from Skardu town. In the meantime, the garrison is running short of supplies. Brigadier Fakir Singh, a senior officer of the Kashmir State Forces which was sent to Skardu sometime back, ran away from there leaving

> behind 65,000 rounds of ammo and large quantities of mortars, hand grenades and rifles. All this has fallen into the hands of the enemy.[43]

Kachru's notes and messages to Nehru during this period form the bulk of communication regarding the complex nature of the Kashmir War and how the final result for either side could not be simply summed up as a victory or defeat. While the Pakistani forces and the tribal infiltrators had been thoroughly routed and their supply lines intercepted, there was certain confusion among the Indian forces too, namely, regarding whether the troops should be moved further and take Gilgit-Baltistan and completely liberate the state of J&K. But there was hesitation in the upper echelons of power in Delhi after Nehru, at Mountbatten's behest, had decided to raise the issue at the UN.

On top of this, there was a mini mutiny brewing within the ranks of the Indian Army, if Kachru's notes are to be believed. In the middle of operations, General Kalwant Singh, who had lionized himself in front of the general public of Kashmir as well as his troops, was suddenly taken off the assignment and posted to Delhi. He claimed it to be a conspiracy against him by two other senior officers, namely, General Lakhindra Singh and Brigadier Usman, who he also claimed had problems with General Thimayya's leadership. These sudden changes in the middle of fiery exchanges between the two forces confused much of the soldiery involved in the fighting, which Pakistan took advantage of.

---

[43]Ibid.

Simultaneously, there was a tussle going on between Nehru and Field Marshal Cariappa regarding whether to induct former INA soldiers into the regular forces. Cariappa was strongly opposed to this, while Nehru and Thimayya were in support. Thimayya, Kachru's notes state, was of the opinion that the irregular forces that had been participating in the bulk of the fighting in the East Punjab region would be demoralized on not being allowed to be part of the regular forces. Ultimately, Cariappa had his way and the irregular forces were not commissioned into the regular army. While it did not breed the kind of discontent within the forces that General Thimayya had feared, it did hamper efforts to lead a coordinated plan forward to retake parts of Jammu and North Kashmir by the army.

5

AFTERMATH

The world has got rid of Hitler and Goebbels, but from what has happened and is happening in my country, I am convinced that their souls have transmigrated to Pakistan.

—Sheikh Abdullah speaking at the United Nations Security Council, 6 February 1948

I am not ready to recognize Pakistan as a party to the Kashmir dispute since they have disowned any responsibility for the invasion of Kashmir and it is fantastic that while the Pakistan delegation to the UN Security Council has refuted the charges saying that the trouble originated from within Kashmir, they are ready to stake anything to get Kashmir.

—Sheikh Abdullah at a reception in Delhi given by prominent Muslim leaders, 17 February 1948

Even after the 1 January 1949 UN-supervised ceasefire, Pakistan did not refrain from its skulduggery. Yes, the Instrument of Accession had been signed by Maharaja Hari Singh. Yes, the promise of a plebiscite had been made by Nehru. But the reality was that the state of J&K that acceded to India was, in large part, never going to be one again. Dangerous tactics were employed by Pakistan to advance their own ends in contravention of the spirit of the 13 August 1948 resolution adopted by the United Nations Commission for India and Pakistan (UNCIP) and the terms of the ceasefire agreement. Border raids continued with a persistence, which at once revealed that they were properly designed and premeditated. The circumstances under which these raids took place, and also details of the consequent loss of life and property were communicated to the Indian Army from time to time by state authorities. The raids were obviously a flagrant breach of the ceasefire agreement and were done with the active connivance of the Pakistan establishment, simply because armed bands could not loiter across what was now the Indian border without revealing their presence to the Pakistan Army, which was manning its own border.

There was deliberate intent behind these continuing raids for they were meant to frustrate the sense of security and peace, which the ceasefire agreement could have ordinarily brought in its wake, and more than that, destroy the possibility of the return of the frightened population of the border areas to their home and hearth. The incursions were an act of war, and it was a systematic attempt all along the northwest and southwest sectors. The idea was to build bases on their side of the border for brewing trouble at opportune moments by smuggling in raiders with the help of locals.

In the areas up in the north, that is Gilgit and Skardu, old revenue records and any other evidence that would establish the fact that Pakistan had been directly governing and administering this area were being destroyed. The records were being replaced by cooked up and fictitious figures and documents to show to the UNCIP that the area, instead of being governed and controlled by Pakistan, as made out by India in its statements, was actually directly governed and controlled by the so-called Azad Kashmir government. Non-Muslims, who were at that time concentrated in various camps in Mirpur, Palandri and Poonch were being subjected to a regular process of victimization and intimidation in order to make them give false statements to the effect that they are being well looked after by Pakistan, and assured that in the future they would be welcome to live under the Pakistan regime. These people were being coerced into making these statements before foreigners and external agencies like the Red Cross. And it was expected that they would be made to repeat this before the UNCIP also.

Significantly, soon after the conclusion of the ceasefire agreement, the government of J&K promulgated an order prohibiting ingress into Indian territories by the so-called nationals of J&K State who had evacuated to Pakistan during the course of the disturbances. The sealing of the border was necessary in order to prevent any influx of people from Pakistan and the Azad Kashmir-held territories. With a view to avoid considerable dislocation in the Indian administrative set-up and to ensure that a suitable identification machinery was set up, no person who was not an approved citizen of the J&K State managed to infiltrate in along with the rest.

Responsibility for sealing the border was fixed on the Indian Army. Despite that, given the porous border, large-scale infiltration from Azad Kashmir areas into Poonch and Mendhar areas was seen. Thinning out of the Indian Army in various sectors as part of the ceasefire was partly responsible for this. Military strategists recommended that the reduction in the strength of the Indian Army at various points should not be considered only in relation to the strategic importance of the place but also in relation to the task of checking infiltration. It was a tough task for a beleaguered army, which had fought incessantly to evict the squatters from Kashmir. A battle none more fearsome than the battle for Kargil in recent times.

Fighting the twin battles on the political and military front, the Government of India received a positive sign from the ruling National Conference and its leader Sheikh Abdullah in the second week of October 1948. At the extraordinary convention of the Jammu and Kashmir National Conference on 13 October 1948 in Srinagar, an 1,800-word resolution of the working committee was ratified, amidst thunderous and prolonged applause, recommending the permanent accession of the state of J&K on the basis of the New Kashmir programme launched in the state. This is the closest the state has come to a plebiscite. Over 250 delegates of all communities from different parts of the state representing popular demand accepted the resolution.

The resolution, which was noteworthy for ratification of the accession, said:

> This convention feels strongly the urgency of convening a Constituent Assembly on the basis of adult franchise for framing a democratic constitution for the State and for giving concrete shape to the aspirations of the people. The delegates expressed their firm belief that Kashmir's future lay only with India and not with Pakistan. This, however, can be possible in an atmosphere of peace and in order to have peaceful conditions in the country liberation of the enemy-occupied territory is the condition pre-requisite. To achieve this most essential objective this Convention urges the Government of India to recognize this necessity and intensify its military activities and bring these hostilities to a speedy and successful conclusion.[44]

It is, however, just as important to examine the Pakistani perspectives regarding the accession, the invasion and the war which followed. Narratives sympathetic to Pakistan show us the diverging realities and the propaganda that was pushed by pro-Pakistani sources to complicate the issue on the international level, now that Nehru had escalated things to the UN.

One such account is by noted historian L.F. Rushbrook Williams in his book *The State of Pakistan*:

---

[44]'Kashmir People Vote for Final Accession to India', *Hindustan Times Archives*, https://tinyurl.com/c9rbmzwm, Accessed on 13 October 2022.

> It seems doubtful in retrospect whether India's action served anyone or anything except her own policy and the Maharaja's waning authority. There were enough Muslim leaders, Muslim troops and Muslim police in Srinagar to protect the city and its inhabitants from the tribesmen, if they had been allowed to do so, but men like Chaudhuri Ghulam Abbas and other leaders of the Muslim Conference—local branch of the Muslim League together with substantial bodies of Muslim troops and police were held in confinement by the Maharaja's orders. But when he himself fled to Jammu on October 26, there was nothing to prevent the organisation of a Kashmir force to defend the city except the presence of Indian troops, who assumed the task of keeping order. Sheikh Abdullah had been let out of prison at the end of September, probably because the Maharaja wanted to gain the support of his NC party.[45]

According to Williams, Sheikh Abdullah promptly renewed his ties with Congress leaders in Delhi, and after the Maharaja's departure, he set up, in the Maharaja's name, an interim administration with India's support, pledged to religious neutralism and non-communal policy on the Indian model. The turn of events took Pakistan wholly by surprise. It appears to them that India, by a sudden coup through the instrumentality of the discredited Maharaja and wholly against the mass opinion of the Kashmiri people,

---

[45]Williams, L.F. Rushbrook, *The State of Pakistan,* Faber and Faber, London, 1962.

set up an administration in Srinagar that was markedly unfriendly to Pakistan, and this, without notification to or consultation with the Pakistan government.

Williams's contention that India never informed Pakistan is a baseless lie just like many other lies and canards spread by Pakistan's political leadership since 1947–48. The Janus-faced Pakistani politicians have never reconciled to losing Kashmir and have been unsuccessful in bridging the trust deficit that exists between the two nations till this day only due to its obsession with Kashmir. Jinnah died persevering that dream, and successive military and political leaders in Pakistan have continued to obsess over poaching Kashmir.

On 28 October 1947 itself, Nehru sent a telegram to Liaquat Ali Khan, saying:

> I have communicated to you text of telegram I sent to PM, United Kingdom regarding Kashmir situation. I have also sent you text of correspondence between Governor General, India, and Maharaja Kashmir, regarding accession of Kashmir State to the Indian Union... I have sent a further message to PM, U.K., informing him of imminent peril of Srinagar and Kashmir from raiders and of action we have taken to give protection to people there. I want to invite your Government's cooperation in stopping these raiders entering Kashmir territory from Pakistan...I wish to assure you that action Govt. of India has taken has been forced upon them by circumstances and imminent and grave danger to Srinagar...In regard to accession also it has been made clear that this is subject to reference to people of State and their decision...Protection of

> Kashmir from armed raids thus becomes first objective and in this we trust we shall have your cooperation.[46]

A second telegram went the same day asking for a meeting with Liaquat Ali Khan at the earliest, and Nehru was happy to note that Khan would be coming to Delhi shortly to attend the Joint Defence Council meeting. Two days later, Khan, in typically Pakistan fashion, replied to Nehru, justifying the acts of violence, loot and arson:

> The position is that Sikh attacks on Muslims in east Punjab in August greatly inflamed feeling throughout Pakistan and it was only with greatest difficulty that Pathan tribes were prevented from entering west Punjab to take revenge on Hindus, and Sikhs. Muslims in Poonch were attacked and those in Jammu massacred by mobs led by Kashmir State Forces and when it was evident that there was to be a repetition in Kashmir of that in East Punjab, it became impossible wholly to prevent tribes from entering Kashmir without using troops which would have created a situation on frontier that might well have got out of control.
>
> Your recent action of sending troops to Kashmir on pretext of accession has made things infinitely worse. The whole of the Frontier is stirring and feeling of resentment among tribes is intense. The responsibility for what is happening is entirely yours...The Pathan raid on Kashmir did not start till October 22nd. It is quite clear, therefore, that Kashmir's plan of asking

[46]Paper on Jammu & Kashmir, Government of India, 1948.

> for Indian troops—and it could hardly have been unilateral—was formed quite independently of this raid, and all evidence and action taken shows it was pre-arranged.[47]

These exchanges of telegrams carried on and are numerous to recount. India and Nehru's tone was always conciliatory, while Liaquat Ali Khan was brash, bold and argumentative. Broadcasts followed from both leaders. Both giving their own version of events—Nehru truthful, Liaquat blasé and brazen. Take Khan's broadcast from Lahore on 4 November 1947, for instance. It showed you a man who was steeped in a mindset of naked obsession for Kashmir:

> In his broadcast, the Prime Minister of India has been tilting at windmills. The armies of Pakistan have not marched into Kashmir as the armies of India, in one guise or another, marched into Junagadh and Manavadar when these States acceded to Pakistan. The Indian Government regarded the accession of Junagadh to Pakistan as a threat to their security. The accession of Kashmir to India is a much greater threat to the security of Pakistan. We do not recognise this accession ... a fraud, perpetrated on the people of Kashmir by its cowardly Ruler with the aggressive help of the India Government.[48]

Still more clarity regarding the sequence of events leading up to the invasion came subsequently when a

[47]Ibid.
[48]Ibid.

controversial letter written by the Sir George Cunningham emerged in mid-1948. The letter was addressed to British commander-in-chief in India Sir Robert Lockhart dated 20 October 1947 warning him of war preparations going on in Pakistan, turned up only in 1948 when Roy Bucher, chief of staff, gave evidence before the UNCIP. The letter contained a warning on an impending attack by Pakistani tribesmen on J&K. Lockhart, it is believed, showed the letter to Sir Roy, but whether this letter was put to Lord Mountbatten or not is not clear.

The record shows that the first communication about the attack of the tribesmen from Pakistan was conveyed to the Indian Cabinet only on 25 October 1947. It was a telegram from the Pakistan Army headquarters stating that some 5,000 tribesmen had attacked and captured Muzaffarabad and Domel, and that considerable tribal reinforcements could be expected. By this time, the invader was literally battering down the gates of Srinagar and but for the heroic defence offered by common people, in turn, helped later by the Indian Army, Kashmir would have fallen in Jinnah's pouch. The fact of the 'Lockhart letter' was, it is pointed out, made known to India only towards the middle of 1948 when the UNCIP started its work in India. Sir Roy, while giving evidence, told the commission about this letter. Surprisingly, when on the commission demanded that the letter be produced, Sir Roy stated that Sir Robert had destroyed it before he left India.

All this time, Pakistan had been denying that she had any control over the raiders or any association with them, while India was trying to convince the world of the fact of Pakistan's participation in the tribal invasion. How such an important piece of evidence in the possession of a top-ranking British officer in the pay of India, which would certainly have used it to establish its case against Pakistan, went missing is something nobody could figure out. The story was first broken by K.N. Bamzai, who had been working at *Blitz,* one of early independent India's most prominent political journals. There was naturally considerable furore in the press and public over this unsavoury episode. Rattled by the story, the Ministry of External Affairs and Commonwealth Relations issued a bold clarification on 6 August 1948, trashing parts of the *Blitz* story:

> Attention of the Govt of India has been drawn to a front page article in a periodical. In this article grave charges have been made against senior British officers of the Indian Army, one of whom retired at the end of 1947 from the post of commander in chief.
>
> It has been alleged that a certain private letter from Sir George Cunningham, then Governor of NWFP to Sir Robert Lockhart was deliberately withheld from the Indian Cabinet and that its contents were recently revealed by the present c-in-c Sir Roy Bucher before the UN Commission in Delhi.
>
> Sir Rob need not have any official use of it whatsoever. Actually he communicated it to the other chiefs of Staff.
>
> Chief of Staff Maj. Gen. Kulwant Singh and then Director Military Intelligence Maj. Gen. Thapar were perfectly aware of the receipt of the letter and its

general contents. No secret was ever made of the matter and it is understood mention was made of it at a Defence Committee meeting.

Further Sir George Cunningham said he doubted if he could stop it. At that time, no denial had come from the Pakistan Government about their complicity in Kashmir, and in fact that the Frontier Province Govt was encouraging the raids appeared to be well known.

Gen Lockhart was sent on a mission to England early in November and returned in the third week of November. He retired towards the end of December and Gen. Bucher became c-in-c from January 1, 1948.

Gen. Bucher has repeatedly mentioned Sir George's letter to high officials - civil and military - in the Ministry of Defence. When he went on the deputation to England in February, he not only referred it in the course of conversation in regard to Kashmir but told Gen. Lockhart he had done so.

The letter was a private communication from Sir George Cunningham to Sir Rob Lockhart. It is quite possible that in those anxious days, when the fate of Kashmir hung in the balance, this fact was not remembered.

In this letter it was stated that Sir George gave a warning of tribal infiltration into Kashmir and that members of the North West Frontier Province Govt were actively helping in this.

The facts make it clear that neither Gen. Lockhart nor Gen. Bucher wished to keep the fact such a letter had been received or its contents a secret

> The information contained in this letter was the first authoritative intimation of the impending trouble in Kashmir and led to the Defence Ministry expedite arrangements for the despatch of arms which had been asked for by the Govt of J&K some weeks earlier.
>
> Almost immediately afterwards came the news of the actual invasion of Kashmir Valley.
>
> While in retrospect, one might regret that the letter wasn't preserved, the Govt are fully satisfied that no one sought to suppress it or conceal it. They consider it most unfortunate that the whole episode should have been invested with sinister motive.[49]

Almost 75 years later, there is no resolution to this dispute over Kashmir, as Pakistan continues to believe that the accession was unsound and a distortion of history. India's position remains constant, and a record of events shows why India is right.

After the issue was internationalized and escalated to the UN by Jawaharlal Nehru on the advice of Governor-General Mountbatten, the organization decided to step in and de-escalate the situation. One of their first attempts at doing so was the McNaughton Plan for demilitarization of Kashmir. In December 1949, the United Nations Security Council (UNSC) entrusted its president, General Andrew McNaughton, to negotiate a demilitarization plan after consulting both India and Pakistan. Pakistan accepted the proposals, but India disagreed to simultaneous demilitarization. The Canadian General's plan looked at

[49]Documents from the private collection of K.N. Bamzai.

demilitarization, preparatory to plebiscite.

The demilitarization programme included the withdrawal of Pakistani forces along with the withdrawal of Indian forces that were not required for the purpose of security, and the reduction, disbanding and disarming of local forces—armed forces and militia of the state and the Azad forces. The Northern Area was also included in the programme, and the two governments were expected to reach an agreement on these proposals by 31 January 1950.

Following the process of demilitarization accomplished to the satisfaction of the UN representative, the plebiscite administrator would then proceed to exercise the functions assigned to him under the terms of the UNCIP for India and Pakistan resolutions of 5 January 1949 and 13 August 1948, as agreed by the two governments.

Nehru's commitment to Kashmir remained steadfast, and the aides-memoires he wrote during this time, summarizing the arguments and countless meetings at the UN with various diplomats confirm this. The first aide-memoire came after Dr Alfred Lozano, his alternate Samper and Dr Erick Colban, the personal representative of the UN secretary general, met Nehru following the formation of the UNCIP and the ceasefire agreement between India and Pakistan, where both Gopalaswami Ayyangar and Sir G.S. Bajpai were also present. Nehru drew attention to Pakistan's repeated acts of aggression against India in the first aide-memoire dated 21 December 1948:

> The Prime Minister informed the Commission that he stood on his original premise that the Pakistan forces must be withdrawn from the State before the

> Govt. of India could consider any further steps...The Commission must realise that there were limits to the forbearance and spirit of conciliation of the Govt of India...Part III of the Commission's Resolution of August 13 provided, 'that the future status of the State of J&K shall be determined in accordance with the will of the people and to that end, upon acceptance of the truce agreement, both governments agree to enter into consultations with the Commission to determine fair and equitable conditions whereby such free expression will be assured.' The present proposals appeared to limit the method of ascertaining the will of the people regarding the future status of the State of J & K to a plebiscite. While the Govt. of India adhered to their position in regard to a plebiscite, they had pointed out that, in view of the difficulties of holding a plebiscite in present conditions in Kashmir, other methods of ascertaining the will of the people should also be explored. The Commission has itself recognised the difficulties of carrying out a plebiscite in Kashmir.[50]

Then, there was a joust over the plebiscite administrator, for Nehru was clear that all that the plebiscite administrator could, in reason, expect was that, for the purpose of organizing and conducting the plebiscite and ensuring its freedom and impartiality, the government of J&K should give him assistance as he might require. There was no question of giving the plebiscite administrator powers

[50]Ibid.

of interference in the administration of the state. The five-member UN commission agreed that their functions and powers would be limited to ensuring that the plebiscite was free and impartial.

Nehru pointed out that the Azad Kashmir forces, which had been armed and equipped by Pakistan and were under the operational command of Pakistan Army, ran into tens of thousands. Their presence in the territories even after demobilization would be a constant threat to the territory under the control of Indian and state forces, a deterrent to the return of many refugees and an obstacle to the free expression of opinion regarding the future status of the state by those who might be opposed to the accession of the state to Pakistan. Dr Lozano, again, confirmed that there would be large-scale disarming of these forces, though it would not be possible to require withdrawal from these territories of genuine inhabitants of these areas.

Nehru then raised the question of the free return to the state of all citizens who had left it on account of disturbances. He added that the tendency of Pakistan would be to push as many people as possible into J&K. And given the length of the frontier and the ease with which people could slip across the border over mountain passes and tracks, a check on the frontier would be neither easy nor effective, except by deployment of a large number of forces.

It is here that the key Kashmir policy mavens redrafted the McNaughton proposal, since the original was unacceptable to the governments, and asked the Indian representatives at the UN to manoeuvre its way with the new plan, which included unconditional assurances to

India from Pakistan that they would deal effectively within their border any possibility of tribal incursion in J&K, and that under no circumstance would tribesmen be able to unlawfully enter the state from or through the territory of Pakistan. Further conditions placed on Pakistan included the complete withdrawal of its troops from Kashmir, along with a large-scale disbandment and disarmament of the so-called Azad forces.

Furthermore, Pakistan had to withdraw its troops from the Northern Area, whose administration would be with the Government of India, subject to such conditions as regards posting of UN observers.

But when this fell through, the great betrayal took place, which only complicated the situation further. On 12 February 1950, Pakistan foreign minister, Zafarullah Khan, literally dropped a bomb during the course of his speech at the UNSC. He admitted that it was upon the recommendation of Pakistan Army Commander-in-Chief General Douglas Gracey that regular units were sent into Kashmir in 1948. Saroja Sundararajan, writing in *Kashmir Crisis*, says, 'Mohammad Ali divulged to Curzon, Member of the British Delegation to the UN, that Gracey warned the Pakistan Government in April 1948 that if it wished Pakistan to survive, it had no option but to send its troops into Kashmir and that Pakistan followed his (Gracy's) advice to safeguard its vital interests.'[51] This disclosure by Zafarullah stunned one and all, and Sir B.N. Rau countered that Gracey

[51]Sundarajan, Saroja, *Kashmir Crisis: Unholy Anglo-Pak Nexus*, Kalpaz Publications, New Delhi, 2010.

was liable to be tried for treason, as he had violated the British Foreign Enlistment Act, which prevented any Briton from engaging in any act of war against a country on friendly terms with Great Britain.

Pakistan's case presented before the UN exposed its deceit, and it offers a perspective on the mindset of Pakistan, which could not accept the fact that Muslim-majority Kashmir in contiguity with Pakistan's borders could actually go to India.

Pakistan's case, based on lies and more lies, and its well-calibrated plan were scuppered by Indian diplomats. UN mediators McNaughton, Owen Dixon, Frank Graham all passed into history, till Nehru and Mohammad Ali Bogra met in London in June 1953 at the Commonwealth Prime Ministers' Conference. Following that meeting, on 20 August 1953, both Nehru and Pakistani Prime Mninster Bogra temporarily agreed to take the issue out of UN hands and resolve it directly. But then Pakistan being Pakistan launched operations in the Rann of Kutch, which spread to Kashmir and Punjab in April 1965.

These military operations and other raids and ambush attacks that Pakistan was able to pull off in a regular manner on Indian soil and in Kashmir was, in large part, possible due to the dithering of the UN and the hesitance of the Indian political establishment to take matters into their own hands and push forward into Gilgit-Baltistan. Intelligence Bureau (IB) Assistant Director M.M.L. Hooja's top-secret report addressed to Ayyangar dated 12 February 1949 provides valuable insights on Pakistan's arrangements for future skirmishes and their stockpiling of weapons.

The report mentioned the construction of ordnance factories at Wah and Sialkot, along with the purchase of equipment from England and the manufacture of weaponry and ammunition at Rawalpindi. Two major bids for the purchase of arms and ammunition were detailed in the report, one known as the Messervey Bid, which took place in January and included the purchase of hand grenades, anti-tank guns, anti-air guns and signal equipment. The second bid, known as the Khanzada or April Bid, concentrated on the purchase of construction and boring equipment for the purpose of being used near the LoC areas to aid in the military's infiltration attempts. Other intelligence obtained from the report included some startling revelations regarding Pakistan's thirst for grabbing Kashmir:

> A New Division has been raised from Rori Division in the name of Bhawalpur State Forces. A scheme for making an arsenal of arms and ammo is in hand. A 150-mile road connecting Hazara with Gilgit is completed. This was a secret high-speed project known as Bamboosa Tract.
>
> Sommer Field track and spare parts of D2 and D4 Tractors were issued from ESSD Jhelum for troops in Kashmir Occupied area. Sargodha refugee camp has been converted into a military camp. Barracks have been constructed and there is concentration of military trucks. Near Chiniot in the lap of the hills, there is more military concentration.
>
> The Pakistan Govt. has obtained services of the sister's son of Pir of Golra Sharif–Chand Badshah–

> to arrange propaganda etc in Pindi district for the forthcoming Kashmir plebiscite.
>
> It is reported that if Pakistan wins the plebiscite in Kashmir, well and good, otherwise Pakistan will fight India. Tribesmen are being sent back home to their homes from the Kashmir front. The forces of Nawab of Amb, Dir and Swat have returned back to their respective States.[52]

This was a comprehensive and overarching view of the enemy by the IB, with sensitive details on how the Pakistan Army was the vanguard of the Kashmir plot. But double-faced as it was, instead of accepting their plotting, the Pakistani leadership chose to attack India with heaps of false propaganda in order to muddy the waters for the UNCIP. In typical fashion, they first launched a broadside at the commission itself. Mind you, this visit by the commission coincided with Jinnah's death, and Pakistan spoke with a forked tongue as always:

> The announcement made by you on the outcome of the recent negotiations has come as a great shock to us. We have still found ray of hope in that part of your communique which stated that you expect to continue your negotiations with both governments 'should you find it desirable', as it showed that the door of a peaceful settlement of the dispute had not been closed. Your reported decision, however, to return to the Continent after the conclusion of your present tour in Kashmir to

[52]Documents form the private collection of K.N. Bamzai.

> submit a report to the Security Council on the result of your labours does, we trust, not indicate any change in the desire originally expressed by you of continuing your conversations with the two Governments, with a view to reaching a just solution to the dispute. If, therefore you still maintain an open mind and are willing to revise the ad interim proposals already formulated by you in the light of the facts that may be placed before you or that may come to your notice in the course of your tour, the present tour will will (repeat) indeed have served a very useful purpose and we shall feel amply rewarded for the very great strain, which it has undoubtedly caused us to record our views in our present state of desolation and grief.[53]

And so, the fulminations carried on as Pakistan heaped abuse on India, the Maharaja and Sheikh Abdullah, making a vengeful case for Kashmir. Take this example from the same letter to the UN where the underlying credo was that it had been cheated out of grasping Kashmir. This memorandum to the UNCIP shows that Pakistan's view on Kashmir remained unwavering despite the passage of time:

> The accession of the State is in the nature of an agreement. A contract whereby each party undertakes certain obligations to the other and acquires certain rights over it. An Instrument of Accession must, therefore, fulfil the conditions required by law in the case of an agreement, if it is a valid document.

[53]Ibid.

> One such essential ingredient is the competence of the contracting parties to enter into that agreement. Could Sheikh Abdullah enter into an agreement with India in this matter? Did he hold a power of attorney from the people or a mandate from them, he could at best speak for his party, which had also not given him any mandate to enter into this agreement. There was a legislature in existence in the State, did it authorise Abdullah or the Maharaja to enter into such an agreement with India?[54]

However, none of these machinations were anything compared to the discontent and confusion brewing in the leadership as well as the general populace of Kashmir, which would ultimately have disastrous effects that are still being felt to this day.

In another undated secret note to PM Nehru, Kachru brings alive the fast-moving shaping of events in Kashmir, which contains yet another bomb primed to detonate:

> Yesterday night Bakshi Saheb telephoned to say that the deliberations of the Security Council were causing wide spread disaffection among the people of the State. The general impression in the minds of the people was that the decision of the UNO would in the long run go against the National Conference and more particularly against the interests of the Hindus.
>
> To this feeling of apprehension has also been added an attitude of helplessness caused mainly by

[54]Ibid.

the devastation and havoc left behind by the raiders. There have been incessant raids and infiltration with consequent havoc and arson on the Sambha–Jammu, about 15 miles away. Hundreds of villages have been burnt down and thousands of families rendered homeless and destitute...The Maharaja has unfortunately got completely demoralised and has been thinking about his own position. Bakshi Saheb was informed of Maharaja's new thinking on the Kashmir issue and was told by Meher Chand Mahajan that a letter to this effect has been sent to the Govt. of India. According to Bakshi Saheb, the latest decision of the Maharaja is that he will intimate the Indian Govt. That his accession to the Indian Union was provisional and that in view of the new development, he had been advised to withdraw himself from this relationship and declare his independence. As an independent Ruler, he would then drive for a military alliance with the Indian Union and try and repel the raiders and secure the independence and integrity of the State.

Bakshi Saheb felt that were the decision of the UNO go against the NC, the leadership would have no other recourse but to side with the Maharaja. He was firmly of the opinion that the NC would under no circumstances yield to any arrangement which would throw them in a position similar to the one in which the North West Frontier Congress was thrown during the plebiscite. They would much rather fight alongside the Maharaja and try to maintain independence rather

> than be annihilated like the Red Shirts in NWFP.
>
> The reports about the proceedings of the Security Council as they have been appearing in the papers in India and on Pakistan Radio are also having an adverse impact on the morale of the Indian troops and officers. The troops are demotivated and disgusted, their thinking predicated on the UN saying that Kashmir State will be forced into Pakistan. Similarly, people in Jammu are fleeing from the State...There is thus a general state of disaffection, disappointment and helplessness. The overall picture is one of a rout.[55]

This note is very dramatic in that it presents a Maharaja of wavering mind after signing the Instrument of Accession. Kashmir was on a razor's edge. In no other integration of a princely state was the path fraught with so many challenges. What comes through from all this documentation is that despite the state's accession, it was poised so delicately that it could have gone any other way, both politically and militarily. While the Government of India was active in its pursuit of a political situation, it arguably failed in reclaiming the J&K State land from the invaders.

[55]Ibid.

6

THE FOREIGN HAND: AMERICANS IN KASHMIR

In the first flush of India becoming independent, Kashmir's geostrategic and geopolitical significance was not lost on the big powers as they positioned themselves in the emerging Cold War game. After the Second World War to overcome fascism, the Allied–Axis powers chose to become pole opposites and bitter adversaries in the reshaping of European borders. Though Kashmir was a fair distance away, diplomats and strategists were convinced of its intrinsic value: it was India's crown jewel, after all, and it touched the international frontiers of many nations. The current United States (US) position could not be more detached from this old stance: in mid-July, the US State Department spokesperson John Kirby ruled out any possibility of the US directly interfering in India's internal affairs; he made a categorical declaration that J&K was India's 'internal matter'.

Some 60 years ago, the US saw Kashmir in a different

light—by the early 1950s, the US had realized that Kashmir should be part of its own sphere of influence. Contrary to popular perception that it was merely an India–Pakistan issue, Kashmir was highly coveted—and not only by Mohammed Ali Jinnah. As it became clearer that the Cold War was only then setting in, the Soviet Union's presence was also growing larger each day in its areas of proximity. Kashmir was seen as one such prime target, given its history of strife and political instability. The US was worried that with Russia flexing its muscles in Azerbaijan, Kashmir was to become the next target for the sheer political vacuum it offered. The growth of communism in the Valley was something that both the Americans and PM Nehru foresaw and were concerned about. It was at this stage that the role of the US began to be monitored in Kashmir.

Even as frenetic dialogue was open between the UNCIP representatives and India and Pakistan to resolve the truce terms, Harry Truman, president of the US, wrote to Nehru on UNCIP and the Kashmir dispute. The message dated 30 August 1949 was delivered to PM Nehru the next day by American Ambassador Loy W. Henderson in Delhi:

> The UN Comm on India and Pakistan has recommended that your Govt and the Govt of (Pakistan) (India) refer to arbitration the determination of truce terms in the State of Jammu and Kashmir. Because of the great importance which my Govt attaches to an early settlement of the Kashmir dispute, I am addressing this message to you in that connection.
>
> Asia is today in the midst of a great social and polit revolution. Unhappily, that revolution has in

most areas been attended by widespread violence which has disrupted the lives of tens of millions of peoples and the economies of whole nations. In the Indian subcontinent, by contrast, essential changes have been and are being made with only localized disturbances, quickly brought under control. It is of crucial importance to the subcontinent, to Asia, and to the world as a whole that this peaceful and orderly progress be maintained.

You will, I am sure, agree that peace and cooperation between your country and (Pakistan) (India) are essential to the continuance of social and political progress in the subcontinent. The Kashmir dispute is the greatest threat to such peace and cooperation. My Govt is, therefore, deeply interested in an early peaceful settlement of this dispute.

Progress toward a peaceful settlement has been impeded during the last several months by the inability of your Govt and the Govt of (Pakistan) (India), with the assistance of the UN Comm, to agree through negot on the terms of a truce in Jammu and Kashmir. Without minimizing the issues which have prevented a negotiated truce agreement, I do venture the opinion that these are not so critical that they should be permitted any longer to impede an advance toward the ultimate goal of an over-all solution of the Kashmir problem. The estab of a truce is in the nature of an intermediate step which will make it possible to give effect to the Commission's Resolution of January 5, 1949.

> In the interest of your own nation and in the interest of the world community, I urge you to accept the recommendation for arbitration of truce terms now placed before your Govt by the UN. I am addressing a similar communication to the Prime Minister of (Pakistan) (India).[56]

While Truman waxed eloquent, he was clearly asking Nehru to accept arbitration under Admiral C.W. Nimitz, the UNCIP-appointed plebiscite administrator, on the truce terms. Beyond that, there was no real support forthcoming to India.

Nehru hit back at Harry Truman after consulting his colleagues and the government of J&K, citing that as the reason in the slight delay in transmitting his reply. He accused Pakistan of double standards and misrepresenting facts on Kashmir, and he made no bones about it. There was sound and fury in Nehru's tone and tenor as he laid out the reality of Kashmir and remained steadfast to his ideal of Kashmir as an integral part of sovereign India. He repeatedly called Pakistan's bluff and blusters in this response to Truman and stated unequivocally that the kind of arbitration prognosticated was unprecedented and as such unacceptable. Both the lawyer and the man of letters are visible in this important missive, where Nehru, despite being forced to take the Kashmir matter to the UN, defended his idea and ideal vehemently:

---

[56]Foreign Relations of the United States: Diplomatic Papers, Government Printing Office, United States, 1977, https://tinyurl.com/7kxcr7yb. Accessed on 13 October 2022.

I greatly appreciate the friendly solicitude which prompted you to write and particularly welcome your frankness because it enables me to write with equal candour. Since India became independent, and the creation of Pakistan was part of the scheme of the transfer of power by Great Britain, we, in India, have been conscious of the need for peace and cooperation between ourselves and Pakistan. One striking example of the genuineness of our desire to be friendly to Pakistan was the transfer to its Government of the equivalent of approximately 150 million dollars at a time when we knew that Pakistan was aiding, abetting the invasion of Jammu and Kashmir. I could cite other instances but don't wish to overweight this letter with detail.

You have referred to the inability of my Government and the Government of Pakistan to agree through negotiations with the assistance of the United Nations Commission on the terms of truce in Jammu and Kashmir. We have, since the Commission first visited this sub-continent last year, given manifold proof of active cooperation with that body. The last example was our readiness to discuss truce terms, under the auspices of the Commission, with representatives of Pakistan. It was a matter of painful surprise to us that the Commission decided to abandon the idea of a conference, in our view, primarily because Pakistan refused even to discuss some matters to which we attach importance, particularly the large-scale disbanding and disarming of the so called 'Azad Kashmir' forces.

About the arbitration Nehru further notes:

> According to our understanding, the Commission's latest proposal as elucidated by the chairman Dr Chyle does not satisfy any of these conditions. On the contrary, the entire work of the Commission and, indeed the Security Council, upto the present is put aside and the problem of the Truce is to be considered entirely de novo, divorced from the moral and practical implications of the developments that have occurred since the dispute started. The Govt. of India maintains that a resort to arbitration in such conditions could serve no useful purpose. It would give no satisfaction to those who like us, firmly believe that acquiescence in the facts of aggression in the interest of expediency beings no lasting solution of a problem but only sows the seed of greater future trouble.
>
> If arbitration is to be just, then due weight must be given to the fact of aggression by Pakistan. The arbitrator must also take into account the interdependence of the phasing of the withdrawal of Indian forces from the State and the large scale disbanding and disarming of the Azad Kashmir forces. Arbitration according to the letter of the Commission's Resolution dated August 13, 1948 and without regard to the consolidation of Pakistan's military position through the strengthening of the Azad Kashmir forces would indeed be no arbitration since a decision on it would avoid the one issue which the Commission has failed to resolve by mediation.

> To conclude, for the reasons given above, the Govt. of India cannot accept the proposal for arbitration presented by the Commission. But as will also be clear from what I have said, they are not opposed to the principle of arbitration. Provision for the settlement of disputes by arbitration exists in the Charter of the UN. As loyal members of the UN, the Govt. of India, always animated by a sincere desire to see a peaceful solution of the Kashmir dispute, are ready to try this method which is novel in the history of the organisation. But the conditions of arbitration must be just and well defined. My Government and I will be happy to consider any suggestion which you Mr President or the Commission may care to put forward in this behalf. Until the basis and terms of arbitration have been satisfactorily settled, we consider it premature to discuss the agency to which arbitration should be entrusted...[57]

The door to arbitration had been slammed shut by Nehru. A position had been taken, and Nehru now stood by it. Over time, between what the raiders took away and what China grabbed, only 45 per cent of the original state of J&K is vested with India.

All this was happening against the backdrop of rapidly unfolding events. On 27 October 1950, the general council of the Jammu and Kashmir National Conference adopted a resolution recommending that a Constituent Assembly be convened to determine the future shape and affiliations of the

[57]Documents from the private collection of K.N. Bamzai.

state of J&K. The area from which this Constituent Assembly was to be elected was only a part of the state of J&K. In a letter dated 14 December 1950 addressed to the president of the UNSC, Sir Muhammad Zafarullah Khan, minister of foreign affairs and commonwealth relations of Pakistan, called attention to Indian press reports that PM Nehru had welcomed the proposed Constituent Assembly and had declared that it would ratify the formal accession of the state to India. Further press reports indicated that a formal proclamation to hold elections was about to be promulgated by the government of Maharaja Hari Singh of the princely state of J&K.

Zafarullah charged that this move sought to nullify the international agreement between India and Pakistan, embodied in the UNCIP resolutions of 13 August 1948 and 5 January 1949. He called for urgent consideration of the Kashmir question by the UNSC and requested the council to call upon India to refrain from proceeding with the proposed Constituent Assembly and from taking any other action that might prejudice the holding of a free and impartial plebiscite.

A covert US intelligence report filed in the summer of 1951 shows how key American officials were involved in the question of Kashmir. The following is an excerpt from the intelligence report:

> I met Frank Collins of the State Department last night. Collins and Howard Meyers used to come from the State Department to New York every time the Security Council met to discuss Kashmir. Along with Mr. Maffet, advisor on Security Council affairs to the

American delegation, they were our contacts on their side. Collins and Maffet were very much alike in their attitude towards Kashmir issue – both pro Pakistan, pro Muslim, pro British and anti Indian ... Maffet being tougher than Collins, but both with a one track mind, unwilling and incapable of understanding points of view other than their own, neither having studied the Kashmir documents carefully.

In contrast, Meyers (Specialist on International Security Affairs, Office of UN Political and Security Affairs) showed some degree of maturity in his knowledge, judgment and approach. Collins arrived suddenly in Delhi on July 5. This morning he left by air for Calcutta where he hopes to spend a couple of days and from where he proposes to return to Bombay via Madras to catch a plane for the US. My plans to have a detailed talk with him were upset by a dinner engagement thrown by Mr. Bourne of the USIS who insisted on both of us attending it. I could not therefore find out with any certainty what had brought him to India apart from the fact that one of his objectives seemed to be to find out how the USIS was operating in India and with what degree of success.

Considering that he arrived in Delhi while Dr. Frank Graham (US Senator appointed Mediator for Kashmir by United Nations) was here, I suspect that his visit may have had something to do with Kashmir also. I commented on the manner in which the Security Council decided to meet suddenly and practically without notice to consider a letter from the Pakistan

> Foreign Minister in which he had taken an exception to the proposal for a Constituent Assembly in Kashmir, the same Security Council ignoring our charges against Pakistan about continuous incitement to war.[58]

To contextualize, it would bear to note what Howard Meyers wrote in a confidential memorandum to the office of the director of United Nations Political and Security Affairs on 3 January 1951:

> Subject: Kashmir Dispute: Possible UK–US Courses of Action.
>
> I discussed the present status of the Kashmir question at some length with Frank Collins, SOA, with particular attention to what we might do if Prime Minister Liaquat of Pakistan does not attend the Commonwealth Prime Ministers Conference in London...
>
> Both Frank Collins and I believe that to do more than this will involve the US taking the initiative in the Kashmir dispute, contrary to our agreement that the British should assume and maintain this initiative. We do not believe that our assumption of initiative is indicated by the present situation. Liaquat's intransigence appears initially, at least, to have strengthened his position at home and that of his government. If the present Western-oriented government of Pakistan should be threatened with dismissal from power because of SC failure to consider

[58]Ibid.

> the Kashmir question and to advance somewhat toward a reasonable solution, then we think the Department should review the situation to decide whether the US should assume the initiative from the UK in attempting to aid the parties to reach a solution...If the UK still refuses to sponsor or co-sponsor a resolution of this nature, we should attempt to secure other co-sponsorship in the SC with the US as one of the sponsors. The UK, presumably, will still co-sponsor a resolution appointing a special representative to interpret the parties commitments and report what has been done in implementation of these commitments. Both Frank and I believe that it is important that the SC go on record as refusing to accept any blatant unilateral attempt to settle the Kashmir dispute, such as the action of the Indian-controlled Kashmir National Conference. We have had some indication recently from New Delhi that the Indians might back down somewhat in regard to this action, if a strong stand is taken by the UN.[59]

According to the above-mentioned intelligence report, the US State Department's Frank Collins had said that despite pressure from Pakistan, and in the belief that India would not welcome an early meeting, the UNSC had kept the Kashmir issue pending between September 1950 and February 1951. As for the talk about 'jehad in Kashmir'

[59]Foreign Relations of the United States: Diplomatic Papers, Government Printing Office, United States, 1951, https://tinyurl.com/bdzzz4v7. Accessed on 13 October 2022.

by Pakistan, the US government had tried, according to Collins, through its diplomatic channels, to bring as much pressure as it possibly could to bear upon Pakistan to favour restraint.

Collins also said that the US still believed in a solution acceptable to both parties, though the US was not willing to leave the problem to the parties for settlement. He referred to Senator Graham—appointed by the UN as mediator—as a sincere man with a high sense of integrity who, with the cooperation of India and Pakistan, 'could help a great deal in holding an early plebiscite.' Parts of another intelligence report said:

> Plebiscite would not be such a difficult thing to hold, I said if the US and UK called upon Pakistan to withdraw their tribesmen, nationals and troops and disband and disarm the Azad Kashmir Force and if the Security Council was willing to honour the assurances that had been given to India in respect of the sovereignty of the J&K Govt and India's responsibility for the defence of the State.
>
> Collins said that under the Resolution of August 13, after Pakistan nationals and tribesmen had withdrawn and Pakistan troops had begun to withdraw, India was to begin the withdrawal of the bulk of her own forces, and the question of disbanding and disarming the Azad Kashmir forces could be taken up only during the Truce period. I replied that such a view was not in conformity with the spirit underlying the Resolutions of August 13, 1948 and January 5, 1949, since the UNCIP itself had admitted in its report that if it

had known that Pakistan would use the intervening period to build up 32 battalions of the Azad Kashmir Forces, it would have dealt with this problem at a much earlier stage.

This being the case, I went on and bearing in mind the fundamental principle on which the August 13 Resolution was based, namely the complete withdrawal from the State of Pakistan troops, nationals and tribesmen, the conclusion was irresistible that the spirit of the Resolution also required the disbanding and disarming of the Azad Kashmir Forces before India could be expected to begin the withdrawal of the bulk of her own forces.

Collins referred to the word 'disposition' of Indian and State Forces in the January 5, 1949 Resolution, whereupon I answered that in the light of the assurances given to India, the word 'disposition' could have only one meaning, namely the disposition of these forces within the State. At this stage the talk was interrupted and we did not get any opportunity to resume our discussion. However, earlier, Collins asked the Indian government functionary [here the intel is referring to K. N. Bamzai, OSD to PM Nehru on Kashmir Affairs] to tell him what he thought of USIS services in India. To illustrate his point, the Indian govt functionary mentioned how the British who had first employed direct methods of publicity in India were gradually forced to abandon them largely on the ground that such methods had failed to prevent the INC from getting stronger and stronger. At this stage

> Mr. Wilkins, first secretary at the American Embassy who was also present turned to Collins and remarked that direct methods were so contrary to US policy that they could never dream of employing them in India or elsewhere.[60]

These reports reveal not only the tensions over Kashmir at the time but also how the US, however distant, was very much invested in the issue. In the aftermath of the Second World War, many western powers like the US viewed Kashmir as the next playground of 'international intrigue', since it was perceived to be the last outpost of anti-communism in the Central and South Asian theatres.

Providing ballast to these postulates was the number-two man of J&K PM Sheikh Abdullah—Bakshi Ghulam Mohammad. Not only did Bakshi remain in constant contact with Delhi through PM Nehru's key intermediary, but he provided critical inputs on the ground situation in Kashmir. In many ways, India's mistrust of the way the UN observers were trying to leverage the role of power blocs in the Valley can be gauged from the note that Deputy PM Bakshi wrote to Nehru on 25 June 1951, only days before the visit of Senator Frank Graham. The note dealt extensively on the role of UN observers:

> It is clear that the UN Observers in the State don't confine themselves to their legitimate function of watching the ceasefire line, but in greater part act as agents of Pakistan. They are obviously abusing their

[60]Documents from the private collection of K.N. Bamzai.

> position as functionaries of UNO and are engaging in what appears to be spying within the State.
>
> It has been gathered that the Observers are chiefly directing their energies towards estimating reactions of people to this government and the accession of the State to India, with a view to drawing inferences about people's real political aspirations.[61]

Then came the bombshell, a revelation so highly serious that it confirmed active US interest and intervention in the Valley at the time:

> Mention may also be made of the activities of the wife of the US Ambassador in India—Mrs. Loy Henderson—who has been in Kashmir for some time. Ostensibly she is here for reason of climate, but it has been noticed that her contacts are mostly with well-known Muslim Leaguers. The inference seems to be more than obvious. Pertinently, Mrs. Henderson even met Sheikh Abdullah during her sojourn in Kashmir.[62]

Earlier, Sheikh Abdullah had had a secret meeting with US Ambassador Loy Henderson in Srinagar, on 29 September 1950, which had been reported to the State Department through a cable. Abdullah was 'vigorous in restating that in his opinion it [Kashmir] should be independent; that overwhelming majority population desired this independence...people had language and cultural background [of] their own. Their Hindus...widely differed

---

[61]Ibid.

[62]Ibid.

from Hindus [in] India, and outlook and background; their Muslims also quite different from Muslims in Pakistan. Fact was that population Kashmir homogeneous in spite of presence of Hindu minority.' But, 'independent Kashmir could exist only in case it had friendship with both of India and Pakistan; in case both these countries had friendly relations with each other.'[63] He even told Henderson that some of the Azad Kashmir leaders favoured this independent status for Kashmir.

Then, in May 1953, Abdullah reiterated this to American Democratic leader Adlai Stevenson in Srinagar on similar lines. Stevenson was on a round-the-world trip after his defeat in the US presidential elections. When Abdullah was finally sacked and arrested, among the charges was collusion with a foreign imperialist power—hinting at a conspiracy with the US to make Kashmir independent of Indian control. But this, too, was contradicted by PM Nehru when he wrote to Vijayalakshmi Pandit, his sister and a diplomat, on 3 October 1953, 'As for Adlai Stevenson, I do not think that he is to blame in any way.'[64]

As if these comments were not damaging enough, particularly the bit about the American Ambassador's wife, Bakshi also forwarded an intercept from Commander John Cadwalader, UNO HQ, Srinagar, to one Dr Francis Fisher Hart in Ambler, Pennsylvania, USA, which is nothing less than explosive:

---

[63]Bamzai, Sandeep, *Bonfire of Kashmiriyat*, Rupa Publications, New Delhi, 2006.

[64]Ibid.

> The work consists in trying by various means to prevent war between India and Pakistan from starting again, and so far this has been accomplished, but because of the stiff necked and uncompromising attitude of Nehru and the Indian govt, I don't know how much longer we can prevail. Incidents keep happening, whereupon we rush to the scene by jeep, horseback or on foot and try to pin the responsibility on someone before retaliation has time to get started. It is sometimes pretty active. Bakshi also sent Nehru an appendix of how for instance UN Observers were taking photographs of Moharra Bridge and river Jhelum. After Abdullah's removal from Kashmir Prime Ministership and his incarceration, Nehru met Pakistan PM Mohd Ali Bogra in Delhi and once again proposed a plebiscite to settle Kashmir with only one conditionality that Admiral Nimitz US envoy to UN not to be made chief plebiscite administrator because of his inherent distrust of super power mentality. But Pakistan insisted and Nehru refused.[65]

Once the Kashmir Assembly under Bakshi's leadership voted for Kashmir's accession to India on 15 February 1954, Nehru concluded that no plebiscite was needed, as the people's representatives had spoken.

That decision of the J&K Constituent Assembly was extraordinarily bold, since it openly defied the then omnipotent UNSC. Some 61 months earlier, the UNSC had been manipulated by scheming Britain into passing a resolution whereby the accession of J&K to either India or

---

[65]Documents from the private collection of K.N. Bamzai.

Pakistan was to be decided only by a plebiscite, although the state had already legally and morally acceded to India through the Instrument of Accession in terms of Britain's India Independence Act, 1947. It is noteworthy that India's case in the UN was predicated on the removal of infiltrators from J&K and not about the legality of the accession. Sheikh Abdullah himself stated at the UNSC's meeting number 241 held on 5 February 1948: 'Whether Kashmir has lawfully acceded to India... is not the point at issue.' He appealed that the UNSC 'should not confuse the issue.' He also reminded the council that when accepting J&K's accession to India, its PM had given the assurance that 'once the country is free from the raiders, marauders and looters, this accession will be subject to ratification by the people.'[66]

That ratification by the people demanded that a constitution be drafted—not by a maharaja (as was done through the Jammu and Kashmir Constitution Act, 1939) but by the people, of the people and for the people. And since Clause 7 of the Instrument of Accession of 1947 did not commit J&K to adhere to the Indian constitution, then being debated in Delhi, a J&K Constituent Assembly was set up through democratic elections, which were witnessed by international observers and conducted on the basis of universal adult franchise—never mind that Sheikh Abdullah's National Conference party was the only one in the fray with its insignificant rival, the Praja Parishad, deciding to stay away.

[66]Chowdhary, Rekha, *Jammu and Kashmir: Politics of Identity and Separatism*, Taylor & Francis, 2015, p. 51.

A flinty Nehru—who always stood by Abdullah the patriot and even banished Maharaja Hari Singh from the state at the behest of Abdullah so that he could have a freer hand while governing J&K—finally had to understand that Sher-i-Kashmir was at best a localist and an inward-looking leader who could not see beyond the narrow prism of Kashmir Valley. Burdened with a Jammu, which he seemingly could not come to grips with, he chose the easier option of engaging with a superpower, namely the US, to exercise his dream of an independent Kashmir, which shared frontiers with five nations as the crown jewel of India. In such a frame of mind, a full integration with India was clearly inimical to Abdullah.

Pertinently, even when Kashmir was needlessly internationalized by India at the behest of Lord Mountbatten, the UN was obsessed with military men to solve the problem. When the ceasefire was instituted on 1 January 1949 after a prolonged war, the UN chose a Belgian, General Maurice Delvoie, as military adviser for the UNCIP. By February, UN Secretary-General Trygve Lie sent 36 military observers—17 from the US, six from Mexico, five from Belgium, four from Canada and four from Norway. In between, there was General Andrew McNaughton, a Canadian, as UNSC president, who dabbled in Kashmir affairs. The original choice for chief plebiscite administrator was General Walter Bedell Smith, who fell ill and was replaced by Fleet Admiral Chester W. Nimitz. Incidentally, General Smith was to later become director of the Central Intelligence Agency.

At one point, even General Dwight D. Eisenhower was in the hunt for the mediator's position, a position which was finally taken up by Senator Frank Graham. The US always subliminally had an interest in Kashmir. Robert Trumbull, writing as far back as 28 October 1947 in *The New York Times*, had said: 'There is sentiment in Ladakh for independence and that such fragmentation of Kashmir is what India wishes to avoid in this strategic state next door to Russia.' The following day, Trumbull wrote something even more damaging: 'Indian officials quite frankly believe that possession of Kashmir is vital to the security of India. Some high government sources interviewed today were thinking about the Soviet Union, which touches Kashmir at Gilgit, at the state's northwest tip.'[67]

Geostrategically, Kashmir's importance is not lost on anyone. While American interest died down after the winding down of the issue at the UN, at the very height of the Cold War, it surfaced in its most vile and obnoxious form. With the 1965 India–Pakistan conflict already a page in history, a fresh American initiative was launched for the state. US State Department troubleshooters were trying to orchestrate a new campaign to test the fertile waters of Kashmir. Known as the Puerto Rican Formula, the State Department began to peddle it as the new elixir.

The US State Department's pointman in New Delhi was Edwin J. Peachaus, second secretary in the US embassy, who sent feelers to top Indian officials between December 1965 and March 1966 about a Puerto Rico-type of commonwealth status for Kashmir to resolve the dispute. His contention

[67]Natarajan, L., *American Shadow over India*, People's Publishing House, 1952.

was that if the 'US could maintain control over Puerto Rico, why shouldn't India follow the same example in Kashmir.'[68]

Ratcheting up the volume, around the same time, *Washington Daily* news correspondent Richard H. Boyce, filed a Srinagar-datelined story on 9 June 1966:

> ...And despite the Tashkent Agreement of last December to settle their differences peacefully, no real progress has been made towards Kashmir peace. Technically India was the aggressor last year. Although Pakistan troops crossed the ceasefire line, Indian troops crossed the international border into Pakistan. Indians think they dare not make any concessions on Kashmir for political reasons at home. The same motivation hardens Pakistan's resolve not to give in. Both governments wrongly tell their voters they won last year's war. Privately, high-level Indian officials confide that they discussed a Puerto Rico-type associate state status for Kashmir—not fully an Indian state, but with some degree of internal self-government though still belonging to India. The hard fact remains that India possesses more than two-thirds of Kashmir and is unwilling to talk to Pakistan about it. India figures neither the US nor Russia will force her to give some of Kashmir to Pakistan. Pakistan is not strong enough to take it by force. But Red China's help could make a difference. China would like Kashmir and it has

[68]Bamzai, Sandeep, 'Retrofit: A Puerto Rico-type option for Kashmir?' *Deccan Chronicle,* 14 September 2016, https://tinyurl.com/2cwnv2f7. Accessed on 13 October 2022.

already infiltrated some of it along the Northeast and Northwest edges. And pro-Peking Indian Communists and Chinese Communists have circulated in disputed areas for months. India resists an internationalisation of Kashmir fearing this would only let China gobble up all of it.[69]

[69]Ibid.

7

THINGS FALL APART: THE KASHMIR PLAN

One of the first salvos that showed the discontent within the political leadership of Kashmir was reported by Dwarkanath Kachru to PM Nehru, and it can be seen, based on his observations, that the PM's man was right regarding his inferences. On 29 March 1948 came the real bombshell, and it dealt with the underlying politics in Kashmir. It was explosive, to say the least, for it provided a peep into what was going to happen in 1953; it was predictive and it came at the wrong time. Kashmir and India were involved in a war of attrition against non-state actors—the UN had entered the equation. On the ground, one needed strong political leadership and, as General Boucher had said, widening and deepening of reforms so that the populace could see the benefits of freedom. Unfortunately, that was not the case as Kashmir and its people lurched between a flawed accession, a tribal incursion and an unending spat between the Maharaja and an emotional Sheikh Abdullah.

Administration suffered, faultiness between the Jammu and Kashmir provinces grew with great rapidity, and Nehru was now preoccupied with affairs of India. Among the myriad problems being faced by the newly acceded state, missiles continued to come in from all directions.

Kachru's note once again highlighted the Dr Jekyll and Mr Hyde persona of Sheikh Abdullah and threw into stark relief his temperament, disposition and demeanour. While Abdullah had delivered Kashmir due to his antipathy for Jinnah and his two-nation theory, his personality being what it was, it was proving to be difficult to deal with his extreme mood swings. A maverick, an ardent agitationist, a staunch nationalist, a brilliant demagogue but a struggling administrator is what Abdullah, Nehru's friend and loyalist, was. His lack of focus and attention to detail was proving a major failing. Kachru's words were harsh about a man who was plagued with self-doubt. Like his notes on the military scenario in Kashmir, his grasp of the political alignments was also very intriguing:

> The differences of opinion between Sheikh Saheb and his deputy Bakshi, resulting the other day in the running away of Bakshi Saheb from Jammu, must not be treated as a mere accident. The differences between the two and in fact between Sheikh Saheb and the rest of his colleagues are many and day by day they are becoming deeper and deeper. These differences are not merely in regard to policy and day to day working of the Government and the conduct of operations but go far deeper. As each day passes by, the differences in outlook and the method of doing

> things or consolidating the political and constitutional gains more and more apparent resulting in frequent clashes and ugly scenes.
>
> Sheikh Saheb has wonderful qualities of head and heart, but he has many failings too. He is extremely honest and sincere and his sense of loyalty towards those who have been associated with him all these years or towards those who have helped him in any shape or form is simply marvellous. But unfortunately he is not a tactician and has no discretion. His attitude towards life and politics is honest, truthful and spontaneous and he often blurts out what he feels...
>
> Having assumed charge of the government and having at the same time to fight the invaders Sheikh Saheb is also feeling rather panicky about the whole situation. Though he is quite confident of his people, he is very sceptical about the outcome of the large-scale ventures launched forth by him and his government. He is obsessed with the idea of the plebiscite and is anxious to avoid doing anything which might in the long run discredit him, politically, and also make it difficult for him to carry the Muslim masses on the side of the Indian Union.[70]

By December 1952, the entire Jammu province was in disarray. Businesses and shops were shut as the clarion call—*Ek desh me do vidhan, do pradhan, do nishan—nahin chalega* (in one nation, two constitutions, two heads of state, and two symbols will not work)—kept the cauldron boiling. Samba,

[70]Documents from the private collection of K.N. Bamzai.

situated on the Pathankot–Jammu road, was the scene of the worst rioting—after all, it was the Dogra epicentre.

The confusion in Abdullah's mind at this stage did not help. With all this drama taking place at the speed of knots, Abdullah was increasingly becoming a prisoner of his own thinking. He was of the opinion that Jammu would never reconcile with the Valley. The seeds of Kashmir's continuing dissent were sown at this time as the religious divide simmered. J&K, from the beginning, did not function as a whole but in parts, and to this day, this situation persists. In a confidential note to PM Nehru after conversing with Abdullah, top government functionaries claimed that he was convinced that Jammu wanted to separate with external support from India. The demons in Sheikh's head were playing tricks with his mind. Rattled, he said the unthinkable:

> It is my definite view that I cannot deliver the goods anymore. Nobody in Jammu is with us and will never be with us. I cannot continue to keep people in the Valley on tenterhooks. So far the people in the Valley have been silent because they know that I will not barter away their interests. Moreover, how can we have peace in the State, if the solution which India and Kashmir adopt is not acceptable to Pakistan?[71]

Abdullah, battling the demons, actually came up short; he was proving to be a narrow-minded leader for Kashmiris

---

[71]Koul, Santosh, et al. (eds), *Kashmir Legal Documents*, Light & Life Publishers, New Delhi, 1977.

and not a universal chief of the state of J&K.

> If today Jammu has rebelled, it is not far off when we will lose all sympathy in the Valley. In the circumstances, independence is best because Pakistan would never agree to a unilateral arrangement and our borders will always be attacked by them. Tell Panditji to have a solution which will be honourable to all—India, Pakistan, Kashmir and him.[72]

Once again, providing an insight into how Nehru had staked everything on Kashmir, he averred:

> Panditji has made this question too personal which it is not. It is a complicated question where so many countries are involved. Panditji has to suffer every time for us. The PM must do this for the land of his forefathers and let him see us settled in peace. I feel we must take advantage of Panditji, he is at the helm of affairs now and has so much influence in the world. UN cannot give us any solution.[73]

When Abdullah was quizzed further on how the logjam could be resolved, he said, 'During the last five years, several solutions have been provided. To me Plebiscite is out of the question. Keeping Jammu people with us is not at all possible. War is ruled out. Let Panditji look at the probable solutions, including that of Jammu and Ladakh (with river Chenab at places as boundary completely merged with India.'[74]

---

[72]Ibid.
[73]Ibid.
[74]Ibid.

The Kashmir Plan was discussed by Sheikh Abdullah and Mirza Beg with a senior government official in Jammu on 20 February 1953. The same was communicated in toto to PM Nehru three days later. It is believed that this plan was a tent pole in convincing Nehru about the prevailing confusion in Abdullah's mind and it accelerated Abdullah's arrest by the director of the IB, B.N. Mullick.

There were other alternatives to the Kashmir Plan that Abdullah was advocating. These were drafted in another note to the PM in April, a copy of which was provided to Delhi by National Conference leader Bakshi Ghulam Mohammad. While some of it may sound like ramblings, the contours of the Kashmir Plan are visible here as well. Titled *Possible Alternatives for an Honourable Settlement between Indian and Pakistan on the Kashmir Issue*[75], it included the following suggestions:

- The people of the State may be given full and unfettered freedom to express their desire whether they wish to acceded to India or Pakistan.
- In order to achieve this objective, the most vital and controversial issue is the conditions under which the wishes of the people will be ascertained.
- The entire State of J&K to be made independent with full guarantee of India and Pakistan as well as the UN.
- The entire State of J&K may be kept under trusteeship of the UN for a period of 10 years after which the people may opt for India or Pakistan.

[75]Vashishth, Satish, *Sheikh Abdullah: Then and Now*, Maulik Sahitya Prakashan, New Delhi, 1968.

- The entire State of J&K under the condominium of India and Pakistan.

These were ludicrous ideas, and Abdullah kept pushing the envelope, saying that failing an agreement on the basis of overall disposition of the state, an alternative mode of arrangements should be considered. It showed deep fissures within the previously united movement led by the National Conference before Independence. Some part of it was due to the misinformation and propaganda being spread by Pakistan and the interference of western powers and the UN, but most of it had to do with the animosity between Sheikh Abdullah and Maharaja Hari Singh, which spilled over and turned into an enmity between the Hindu-dominated Jammu and the Muslim-majority Valley. Another set of equally incredulous alternatives, which showcased the extent to which Abdullah's thinking had departed from that of the Indian political leadership, suggested by the National Conference were:

- The Hindu-majority areas in the south may go to India, the Muslim-majority areas in the northwest to Pakistan, and the Kashmiri-speaking areas may be held under the condominium of India and Pakistan.
- The state may be divided into different zones. The Hindu-majority areas in the south may go to India and the Muslim majority areas in the northwest may go to Pakistan. The Kashmiri-speaking areas may be demarcated with due regard to viability and security of this unit, and the wishes of the people in this zone may be ascertained through a plebiscite.

- The Hindu-majority areas in the south may go to India, the Muslim-majority areas in the northwest to Pakistan, and the Kashmiri-speaking areas may be formed into an international unit with treaty relations with India and Pakistan.
- The state may be partitioned on the basis of religion, and the Hindu-majority areas may be transferred to India, and the Muslim-majority areas to Pakistan.
- The present ceasefire line may be established permanently with suitable adjustments.

Abdullah was, by now, obsessed with his insular and myopic view of an independent Kashmir Valley, which he could preside over, not caring two hoots about the rest of the truncated state left with India. The entire doctrine went against the grain and was a volte face from the man who had opposed Jinnah for his two-nation theory built on the edifice of religious identity and division. It sent alarm bells ringing.

Then came the proverbial straw that broke the camel's back—Abdullah writing to Nehru on 4 July 1953:

> I realise that you must be having an unusually crowded time after your return to Delhi. I would have certainly come down myself to Delhi, but Bakshi and Beg will explain to you why it is not desirable for me to leave Srinagar at present. I would very much wish you to come to this place after the AICC session is over. Maulana Saheb (Azad) could also come with you. In any case, it would be useful that we meet here before you visit Karachi. Objectively the State is subject to pulls from India and Pakistan. The external pressure

> naturally creates internal reactions in terms of divided loyalties. In order to neutralise these reactions, we had devised a formula and considered a restricted relationship with India as a suitable course conducive to internal consolidation. We believe that the accession of the State to India on the terms of the Instrument of Accession would provide necessary opportunities of allaying the fears of various sections of the people of the State.[76]

And so, the emotional outpourings of a disturbed mind continued in the lengthy letter:

> When Article 370 was devised, we felt assured by the statement from Sardar Patel that the Instrument of Accession would be the final basis of India–Kashmir relationship. Subsequently, when the Delhi Agreement came up before the Council of State on August 5, 1952, Gopalaswami Ayyangar stated that Article 370 was not a permanent feature of the Indian Constitution and 'when the time was ripe', this provision could be removed from the Constitution.

The litany of woes continued till he ended with a sense of victimhood, 'My idea about secular democracy is not cramped or narrow-minded, I believe in justice for all sections of the people and my attitude is conditioned by realities and by wishful thinking. I agree that personal relationship between individuals should not be a

[76]Documents for the the private collection of K.N. Bamzai.

consideration where larger national interests are involved.'[77]

In this battle between the road to integration and greater autonomy, the die had been cast by both sides. The split was defined and wide open. Not just between Nehru and Abdullah, but the cleave was now drawn within the National Conference, where Bakshi Ghulam Mohammad was openly batting for India, while Mirza Beg was batting for Abdullah. In an internal National Conference coup, Abdullah was toppled by Bakshi, who took over the reins of the state as its PM.

The growing estrangement between Sheikh Abdullah and Pandit Nehru was one of the most unfortunate happenings in the years that followed Independence. Kashmir's accession, though controversial; Lord Mountbatten's unfortunate advice to go to the UN, something that Nehru regretted; and, finally, Abdullah's arrest and Bakshi's installation—the wheel had come full circle in a short span of six roller-coaster years. The ever-growing Jammu–Kashmir divide had taken its toll, and the biggest casualty was Abdullah himself. He was someone who had never reconciled to Jammu, for he regarded it as a symbol of the Dogra dynasty.

Finally, on 8 August 1953, Nehru gave the orders for Sheikh Abdullah's dismissal as the PM of J&K by the Sadr-e-Riyasat, Dr Karan Singh. Soon after, Abdullah and 23 of his supporters from the National Conference, including his biggest cheerleader, Mirza Beg, were arrested on the grounds of 'anti-national' activities.

The government of J&K opened a legal case against the 24 members of the National Conference in what has

[77]Ibid.

become known as the infamous 'Kashmir Conspiracy Case'. The charge of 'collaborating with Pakistan', a ludicrous charge, considering all that the National Conference and Abdullah, in particular, had done for the accession of the state to India. However, it could not be denied that the confusion and indecision of the state leadership as well as the Indian leadership coupled with the constant interference of Pakistan and the US had led to this.

Abdullah and his supporters spent almost 11 years in jail, with the court case carrying on till 1964, when Nehru decided to revoke the case shortly before his death.

EPILOGUE

Recently, the Modi government has tried to integrate J&K with the rest of the country; it has built infrastructure and attempted developmental initiatives rapidly after the revocation of Article 370 without ever compromising on national security or its Westphalian hegemony over the Valley. The tortuous and often treacherous pathways of the past, which always led to a cloud of flies buzzing over India's constitutional, legal and moral right over the Valley, have been consigned, hopefully, to the rubbish heap of history. Yes, Pakistan and its dreaded S Wing refuse to accept the new reality and continue their stratagem of death by a thousand cuts, but even within the Valley, there is fatigue to this theme that thrives on bloodletting.

The contentious years that resulted in the making and unmaking of the Kashmir problem have been dealt with through a brand new prism using the help of documents and correspondence hitherto not in public domain. Kashmir turned into a brutal caged match, a diabolical kill box, where Pakistan, in trying to mask the scent of the snare, has and is creating consternation in the Valley.

From Mohammed Ali Jinnah's naked obsession for

Kashmir to Sheikh Abdullah's staunch hatred for Jinnah and his Islamic ideological identity politics to Jawaharlal Nehru's comprehension of the Valley being a shop window for his brand of secular politics and hence his unshakeable idea that the princely states fit into the bigger ideal and idea of a Union of India to Maharaja Hari Singh's streak of remaining independent in the face of both India and Pakistan, this treatise covers the vital years that defined Kashmir's accession to India.

The role played by two young Kashmiris—Dwarkanath Kachru and Kashi Nath Bamzai—in helping Nehru close the Kashmir accession cannot be underscored. Nor can V.P. Menon or Sardar Patel's role in this closure be ignored. His famous words in the Defence Committee meeting as Nehru vacillated—*Jawahar, Kashmir chahiye ki nahin* (Jawahar, do you want Kashmir or not)?—continue to reverberate in the corridors of my mind. Sheikh Abdullah sitting in the anteroom; Jinnah calling up endlessly with his snake oil pitch to speak to Sheikh; Nehru, Sardar and Baldev Singh waiting for Menon; and Sam Manekshaw returning with the signed Instrument of Accession; the raiders breathing down on Srinagar—it was all touch and go and nothing short of high-decibel screenplay of a potboiler.

Rewind to 8 December 1989, after much tumult and controversy, with a V.P. Singh National Front government recently in place, a momentous event takes place. Union Home Minister Mufti Mohammad Sayeed's daughter Rubaiya is abducted by Jammu and Kashmir Liberation Front (JKLF) militants, and the world is turned upside down. Top erstwhile Jan Morcha leaders arrive at the house of Sayeed, where he is inconsolable as Arun Nehru, Arif Mohammad Khan and Satya Pal Malik (governor during this period) are trying to convince him to appear on national television to say that she is the nation's daughter and it's imperative that she be set free. But a tearful Sayeed has lost all reason to think, overcome by extreme emotion, for obvious reasons. He refuses to do anything—stunned and struck by inertia.

Just six days earlier, Mufti had taken oath as the first Kashmiri Muslim home minister in V.P. Singh's government. At the same time, JKLF's Asfaq Majid Wani wanted to do something spectacular in Kashmir Valley. His charter was to kick-start the 'revolution', and he didn't know where to begin. Watching the oath-taking of Mufti, he thought of an audacious PLO-type plan to rattle the newly formed government. The original plot, conceived by Wani, was to kidnap Mufti's son, reportedly a doctor in Lal Ded hospital. But once recce was carried out, the son turned out to be a daughter—Dr Rubaiya Sayeed. As she finished her shift and left for home around 3.00 p.m. on 8 December, boarding a bus at Exhibition Crossing, JKLF militants took over the bus, with Wani and others following in a car.

Around 5.30 p.m., JKLF top brass, Javed Mir, called up *Kashmir Times* and relayed the news of the abduction of the home minister's daughter. All hell broke loose, with phones ringing non-stop in the Valley and Delhi. The triumph of V.P. Singh slaying Rajiv Gandhi was lost in translation as panic gripped the security mavens.

After 122 hours in captivity, against the wishes of J&K Chief Minister Farooq Abdullah, five top separatists were released for Rubaiya. It became a watershed moment for Kashmiri militants, as they brought India to its knees. JKLF satraps Wani and Yasin Malik had succeeded in delivering a killer blow to Kashmiriyat.

Fast forward to 18–19 January, and the Kashmiri Pandit stampede was set in motion to complete the construct of a Muslim Kashmir Valley, a gambit that the JKLF was propounding and working on.

Jacked in and booted up, resilient Kashmiri migrants still long to return to their homeland. Given that the arc of a moral universe is long, perhaps a day will come when they can put recriminations aside and beam themselves, most unlikely at the moment though. It is now 32 years and more since they were stampeded out of their home and hearth in the Valley. The patina of a bygone era continues to haunt their brutalized psyches, as they remember the infamous date—19 January 1990. Several crucial instances of targeted ethnic cleansing were used to create a fear psychosis, the brute majority swelled on to the streets shouting slogans. The sloganeering was deafening yet focussed, for there was a strategic imperative behind it. The clear and present danger was to the Pandits, and the motive was to chase them out. Fierce cries reverberated: '*Hum kya chahte—Azadi. Azadi ka matlab kya—La ilaha illlala* (What do we want—freedom. What does freedom mean—No one but Allah)'; '*Yahan kya chalega—Nizam-e-Mustafa* (What will rule here—Islamic law)'; '*Asyi gasyi Kasheer, batav warai, batnyav saan* (We want Kashmir, without Kashmiri Hindu men, but with Kashmiri Hindu women)'—and many more! Who would tolerate this and that too in a place where the minority community had had, till then, the best of relations with majoritarian rule? The brutal ethnocentrism had shattered the glory of Kashmiriyat once and for all.

A fellow Pandit likened us to aborigines the other day, and I thought that description was apt, for we are migrants and nomads in our own country. This can only be described as a monumental travesty. Over time, Kashmiri Pandits have vacated their space in the Valley and left for the plains, but this was the seventh and final exodus. It started on 14 September 1989, leaving only a few thousand of them behind in the Valley, and about 4.5 lakh Kashmiri Pandits were left to live the life of refugees in their own country.

The act of losing something can ensure that you elicit great wisdom. For the hapless Pandit, this wisdom translates into pain. For how do you think it feels if you are displaced in your own country, ignored by different political parties and in many ways expunged from the national discourse? The departure from Kashmir and lack of rehabilitation constitutes one of the greatest human tragedies since Partition. The Pandits, today, are a lost community, and perhaps at one level, a lost generation. It is a tribute to their workman-like nature that they have picked up the pieces and gone back to leading their lives normally, all the while suffering from PTSD and other related trauma issues.

Now, years later, many such Kashmiris are still struggling to come to terms with their displacement, trying to subsist in alien conditions. Though, it is not the same thing, at least the majority have thrown themselves back into the hurly burly of a daily regimen. Survival, after all, is at the kernel of life, and since nature abhors a vacuum, life has to go on.

The community, meanwhile, deprived of their *madre vatan* (mother land), still makes its presence felt.

I must confess I am not a refugee, though I was born in the land of my ancestors, very much in the vale of Kashmir. Against that, I would like to believe that I am an Indian first, having lived and worked in three of the biggest metropolitan cities of this wonderful country—Delhi, Kolkata, Mumbai—and now, once again, in the nation's capital. But at another level, I am also a Kashmiri; my roots call out to me constantly. They ask me searching questions, they probe, they irritate my subconscious, wanting to know what I have done about my heritage, legacy, call it what you will. But let's go back to the stampede of the Pandits. It didn't happen overnight. The Pandits had lived through other such attempts in the past. What changed? The systematic weaponized pogrom to target all those who symbolized 'India' in the Valley worked like clockwork. The terror network thrives on creating a fear factor. If that meant going after co-religionists, then so be it.

A case in point was a National Conference worker and political activist Yusuf Halwai, who had illuminated his sweet shop on 15 August to celebrate India's independence. This perceived 'outrage' saw him being slain and a message sent to the majority community in the Valley that India was a pariah, and anyone who had anything to do with India, would be treated like Yusuf Halwai. That was phase one—create a fear psychosis amongst the majority community.

In Anantnag, another advocate and activist, Prem Nath Bhat, met the same fate. But the breaking point came when a nurse, Sarla Bhat, was abducted from the Sher-e-Kashmir

Medical Institute in Sohra, gang-raped and then her mutilated body thrown into the Habbakadal thoroughfare.

In yet another killing, Bal Kishen Ganjoo was done to death in his own house. The message for the minority community was loud and clear; in fact, it was a neon sign being held to their faces. 'Get out' is what it said. Many of my relatives, including my mother's sisters, had to flee the Valley in the darkness of the night. Now let me inject a hard dose of reality.

Between December 1989 and 15 May 1990, the J&K government records show that 134 innocent persons were killed by the rampaging militants. Amar Kapur, additional DGP, Kashmir, of the time, said: 'There is no truth that the state government extended help to the Pandits to leave the Valley.' The stampede was reportedly hastened by the state administration. Jagmohan refutes the charge levelled by People's Union of Civil Liberties: 'With regard to migration of non-Muslims (KPs) from the Valley, I had stoutly challenged the assertion that the state government had provoked or sponsored it. When they (PUCL) suggested that they have come across use of government vehicles for the purpose, I told them that individual misconduct by a government servant in the use of a government vehicle in his custody cannot be ruled out in such a situation.' Farooq Abdullah quit on 18 January; in the intervening night of 18/19, the stampede took place; Jagmohan arrived as governor the following day. The horse had bolted, all was lost, Kashmiriyat was dead, its bonfire lit by Islamists.

Following the Rubaiya playbook, innumerable abductions took place, and the release of many militants took place. The period between December 1989 and January 1992 saw frenetic abductions.

Prominent among them was the abduction of Tassaduq Dev, brother-in-law of then Union Parliamentary Affairs Minister Ghulam Nabi Azad by Al Umar Mujahideen in January 1992. Three jailed activists of Al Umar were set free in exchange for Dev's release on 17 January. Before this, came the abduction of Nahida Soz, daughter of then National Conference MP Saifuddin Soz, by the Jammu and Kashmir Students Liberation Front (JKSLF) in August 1991. Nahida was released when the government set free a Pakistan-trained hardcore militant, Mushtaq Ahmed.

Indian Oil Corporation's executive director, K. Dorraiswamy, was abducted by activists of Ikhwan-Ul Muslimeen in Srinagar on 29 July 1991. His release on 21 August was possible when the government set free six militants. The released militants included Javed Shalla, main accused in the kidnap and murder cases of Kashmir University vice-chancellor Mushir-ul Haq and HMT general manager, H.L. Khera, in 1990. Seven more militants were set free to seek the release of Mir Nassar Ullah, son of former J&K minister G.M. Mir Lasjan; A.K. Dhar, scientist; and T.K. Raina, retired deputy commissioner, were abducted in March 1992.

The Rubaiya Sayeed case had set a precedent for kidnappings for seeking release of jailed militants. According to government statistics, the state witnessed an upsurge in abductions after Rubaiya's kidnapping. While only one kidnapping, that of Rubaiya, was reported in 1989, 169 abductions were reported in 1990, 290 in 1991, 281 in

1992, 349 in 1993 and 368 in 1995.[78] It virtually became a cottage industry.

Incidentally, in one of these kidnapping cases, no militant was released for seeking the release of Mushir-ul Haq and H.L. Khera in 1990. They were killed by the captors.

These were gravity-defying times in Eastern Europe and Soviet Union; there was unrest and rebellion, nations were being mowed down, their complexion and character undergoing metamorphosis. Those orchestrating these moves in Kashmir and their handlers in Pakistan probably had this misconceived notion that perestroika would be unleashed in the Valley too. The radio and TV stations would be captured, and the umbilical cord with Hindustan would be cut forever. But they had not reckoned with the resolve of the Indian state, which, while accepting the stampede of the Kashmiri Pandit, did not allow anything untoward to happen. It resisted the glasnost and retained the Valley, but the Pandits became collateral damage.

Kashmir remains a festering sore, and while much blood has flown over, and it is relatively stable now, the Kashmiri Pandit has been erased from the state. Over these years, the terror network has upped the ante by indulging in random gruesome massacres of Kashmiri Hindus, serially in Nadimarg in 2003, Wandhama in 1998 and Sangrampura in 1997. The circle of terror was complete. Fear had transcended the vale.

Years later, all of us, in our own small ways, grapple

---

[78]'Terror Trail-V: The politics of abduction in Kashmir (IANS Exclusive)', *Business Standard*, 11 May 2019, https://tinyurl.com/3xw85a9r. Accessed on 13 October 2022.

with this reality of being taken out of the Valley. I cannot imagine how those who had to migrate out of the Valley coped with this harsh reality, how they dealt with the demons in their heads, how they adjusted to their new locations and environment. I know several of my relatives, both paternal and maternal, have rebuilt their lives. The process has been difficult, but they were economically better off than a whole host of families. But they, too, seek a tryst with their homeland, just as I do.

Weevils in the flour sack have derailed the Kashmir story—political families controlling the state-nurtured corruption, while funambulists in Delhi walk the fine line. The 'Kashmir Factory' has remained a cash-guzzling machine as separatists stoked the fires with the carefully calibrated and ratcheted lingua franca of azaadi. This epistemology of Kashmir, where syncretic faith, belief and culture was brutally disrupted by weaponized Islamic ideology, remains an unending travesty.

INDEX